T0314232

THE WASTE CRISIS

THE WASTE CRISIS

Roadmap for Sustainable Waste Management in Developing Countries

SAHADAT HOSSAIN

University of Texas Arlington, Arlington, TX, USA

H. JAMES LAW

SCS Engineers, Raleigh, NC, USA

ARAYA ASFAW

Addis Ababa, Ethiopia

Registered Office(s)
John Wiley & Sons, Inc., 111 River Street, Hoboken, NJ 07030, USA
John Wiley & Sons Ltd, The Atrium, Southern Gate, Chichester, West Sussex, PO19 8SQ, UK

Editorial Office
The Atrium, Southern Gate, Chichester, West Sussex, PO19 8SQ, UK

For details of our global editorial offices, customer services, and more information about Wiley products visit us at www.wiley.com.

Wiley also publishes its books in a variety of electronic formats and by print-on-demand. Some content that appears in standard print versions of this book may not be available in other formats.

Library of Congress Cataloging-in-Publication Data applied for
Hardback: 9781119811930

Cover Design: Wiley
Cover Image: © Unsplash/CC0 Public Domain

Set in 9/13pt Ubuntu Regular by SPi Global, Pondicherry, India
Printed and bound by CPI Group (UK) Ltd, Croydon, CR0 4YY

C9781119811930_260722

Contents

Preface

The world is rapidly moving toward urbanization, and the amount of municipal solid waste (MSW), one of the most important by-products of an urban lifestyle, is growing even faster. Urbanization or "urban transition" is a process of shift in population densities from a rural/agriculture-based economy to a denser population with an industrial and service-based economy. Urbanization has generally been a positive force for economic growth, poverty reduction, and human development. With higher percentages of young urban dwellers, economic activity increases significantly in urban areas, which contributes to GDP growth of urban population in a country. Increase in population in urban areas and GDP growth results in increase in consumption ultimately leading to an increase in waste generation. Therefore, based on the latest data available, global waste generation in 2016 was estimated to have reached 2.01 billion tons (4.4 trillion lb) (World Bank Report 2018a). By 2030, the world is expected to generate 2.59 billion tons (5.71 trillion lb) of waste annually, and by 2050, waste generation across the world is expected to reach 3.40 billion tons (7.5 trillion lb).

However, sustainable waste management is a major issue for both developed and developing countries. Traditionally, as solid waste management practices and collection improve, waste will begin to be placed in open dumps or landfills, which still poses human health/safety threats via disease vectors, water pollution, and explosive conditions. Even though, source reduction or recycling is preferred choice of waste management, over 70% solid waste is open dumped or landfilled globally. Locally and globally, existing solid waste management/mismanagement presents challenges and opportunities.

In developing countries some current waste crises are presented as follows:

1. 1/3 to 2/3 of the solid waste is currently dumped indiscriminately in streets and drains, where it can breed insects and rodents that transmit communicable diseases such as dysentery, typhoid fever, cholera, yellow fever, and plague (Zhu et al. 2007).

2. Waste is traditionally burned in the streets and fields in and around city centers, producing dioxin emissions and damaging the air quality through soot emissions and the climate through carbon dioxide emissions. Poorly managed waste has an enormous impact on health, the local and global environment, and the economy, and improperly managed waste usually results in down-stream costs that are higher than the cost of managing the waste properly in the first place.

3. In developing countries, the rate of waste collection is between 20% and 60%. Uncollected waste dumped in open places, on roads and streets, in water bodies, and in most public areas is getting into city's drainage system. Presence of highly nondegradable plastic bottles and plastic bags are clogging the city's drainage system and causing flash flooding.

4. Inappropriate operational practices are causing serious health hazards for waste pickers at working face of a dumpsite. In some cases, failure of an open dumpsite is costing hundreds of people's lives living next to an open dumpsite (as in Koshe Dumpsite in Ethiopia).

5. Recent China Ban on importing recycled products from global market has caused serious effect on recyclables plastics and other waste components. Many cities are discarding their recycling programs and sending them to landfill as they did not have a local market for reusing the recyclables. Years of hard work to build recycling system and market needs revival with incentives from city officials or regulators. The problems created by China Ban can be a great opportunity for local young entrepreneurs to create new products and market using recyclables.

Regions of Africa, Latin America, South Asia, East Asia, the Pacific Islands, and the Caribbean are urgently in need of help in mitigating problems associated with increasing population growth, urban consumption, and waste production. If safe waste management practices are not incorporated into their development plans, the proliferation of poor health and sanitation conditions will persist. The real challenges of successful solid waste management are essentially those listed below.

- Lack of public awareness of waste management and its impact on health and well-being, the environment, and the local and global economies.
- Insufficient collection and management systems in developing countries that cause major health, sanitation, and environmental issues.
- Improperly managed open dumpsites and their serious consequence of environment and public health.
- Lack of space for new landfills every 20/25 years as urbanization and migration of population to urban areas causes serious strain on urban waste management system.
- Inadequate sustainable waste management technology and implementation of the technology in developing countries.
- Lack of training on waste management technology and lack of management assistance for local authorities and waste management personnel in developing countries.

- **The marginal or nonexistent guidelines or ROADMAP** for developing waste management protocols or regulatory frameworks at both the local and national levels in many of the poorest countries in the world.

In Developed Countries: (i) Landfills typically occupy an area from several to hundreds of acres, and the current lack of available space for new landfills is a real problem for future waste management. Due to rapid growth and urbanization of cities beyond their current limits, many previously closed landfills which were outside the city limit during their closure, are now within the city limits. Opening a new landfill within the city limits often causes violent protests similar to those which occurred when demonstrations and protests against the opening of a new dump (landfill) on the slopes of Mount Vesuvius in October 2010 led to a riot and violent clash between the local residents and police. (ii) Waste minimization and the reuse of existing landfills is key for sustainable urban development, but even the most preferred choices of waste management, recycling, and reuse have inherent problems. For instance, Sweden is a recycling-happy land. The incineration of solid waste provides power to 250 000 of their homes and heats 810 000 homes. Recycling is so effective in Sweden that only four percent of all waste generated in the country is landfilled. However, the Swedes ran out of garbage needed for the incinerators and have had to import it from Norway (Hickman 2018).

In the last 40–50 years, developed countries have moved from open dumps to sanitary or engineered landfills, while the developing world is still practicing open dump methods. Very recently, the developing countries in South Asia, Latin America, Africa, and Eastern Europe have begun transitioning from open dumps to sanitary landfilling systems. However, based on the increasing rate of urbanization and the rate of increase in waste generation, the current ways of managing waste through landfilling or sanitary landfill systems may not be sustainable. Because of urbanization, it is almost impossible to find land on which to build new sanitary landfills every 20–25 years, so that is not a sustainable solution. The problems associated with the waste generation and management by current systems are summarized below.

- Populations in urban mega cities are increasing at an alarming rate, and the amount of waste that is being generated is also increasing.

- The available land for building residential houses to accommodate the influx of people from rural areas to urban areas is shrinking every day. Consequently, the cost of land for building a landfill within the city or in nearby cities is extremely high.

- The level of resistance to building new waste management facilities (landfills) within communities is very high, which causes major distress within the cities' political system. Therefore, the policy makers or city officials are reluctant to approve the construction of new landfills within the city limits.

- Waste-to-energy (WTE) or incineration can provide a lucrative solution to waste management as WTE plants can address the issue of land/space in both developed and developing countries. It is an appropriate technology for waste management in developed countries such as Europe, the USA, Japan, and South Korea, where the waste is relatively dry and the substantial presence of plastics, paper, and wood make it a good source of combustible materials. The waste in developing countries has a higher percent of food waste (more than 70%) and high moisture content than that of the waste in European countries. Therefore, generating power through WTE or incineration, using this low calorific (organic waste) and highly wet waste may not be applicable or cost-effective for developing countries, if without detailed studies and analysis.

The hierarchy of waste management may not be the same for all countries or all kinds of solid waste generated. In other words, **ONE SOLUTION DOES NOT FIT ALL**. The handling of solid waste requires solutions that are flexible, but robust enough for urban sustainability.

A clear understanding of waste generation, collection, and management practices, as well as **a roadmap for sustainable waste management in developing countries is vital** for environmental sustainability, good health, the safety of waste pickers and people living in communities near dumpsites, and above all, for creating healthy urban cities across the globe.

The proposed sustainable **Resource Management** system, **S**ustainable **M**aterial **A**nd **R**esource **T**reatment **(SMART)** facility, would replace traditional landfills and open dumps and be a viable solution for developed and developing countries. SMART facility will eliminate the problems associated with loss of materials, climate change impacts, post-closure monitoring costs, and most importantly the major roadblock for sustainable waste management **SPACE**. It will facilitate greater material recovery and reuse, accelerate waste degradation rates and renewable energy generation, perpetuate operations in the same location, improve the public's perception, create jobs, and enhance the acceptance by the greater urban community. Once source-separated, mixed waste can be processed in a SMART facility, and sustainable material/resource recovery and management will be a ONE-STOP operation. The design and operation of conventional landfills will be replaced by a biocell (within SMART facility). Biocell acts as a perpetual landfill and is a major part of a sustainable waste management system in which waste is never landfilled permanently. Rather than serving as a permanent storage facility, the biocell is a temporary repository that is used to retrieve all the potential benefits during the repository period. It will also accelerate waste decomposition and replace landfilling by treating the waste so that it produces greater levels of renewable energy quickly.

The proposed sustainable solution covers all the major components of sustainability, as well as the technical, economic, and social aspects of the system. The government shoulders the responsibility of collecting and managing waste. Consequently, significant portion of cities budget is allocated to municipalities waste management. The challenge is to shift the burden from the city to the private sector by creating or enabling an environment for the business to flourish throughout the value chain from collection to processing waste to valuable products which are used by consumers while creating sustainable employment for urban dwellers with living income. Therefore, in the framework of Public Private Partnership (PPP) the SMART facility should be established to accommodate the interest of the government, the private sector, and the public at large. Therefore, SMART facility has the potential of creating of jobs for the people living around the waste management facilities, making it a source of income rather than a source of distress, sickness, and economic hardship. If the waste management system is managed correctly and sustainably, the unhealthy conditions that are presently observed all over the world can be minimized significantly and it should provide healthy living conditions for both poor and rich residents.

Developing countries need help from advanced or developed countries for appropriate technology of managing their waste. What works in one place/city/country may or may not work in another country. Therefore, training and human capacity building is vital and important component of sustainable waste management. Many developed countries are willing to share their experiences and offer their tried and tested solutions but finding an effective regional platform to deliver this exchange is simply absent.

The Organized Research Center of Excellence (ORCE) – Solid Waste Institute for Sustainability (SWIS) founded in January 2015 at the University of Texas at Arlington (UTA) is trying to bridge the gap that currently exists between developed and developing countries. The international collaborative partner of SWIS for training and capacity building is the International Solid Waste Association (ISWA). **The mission** of SWIS is to develop clean and healthy urban cities through sustainable waste management.

SWIS conducted its first international waste management training through ISWA-SWIS Winter School in January 2016. The training program is a collaborative partnership between ISWA and SWIS. The objective of the ISWA – SWIS Winter School was to provide advanced knowledge in the field of waste management to an international audience of existing and emerging solid waste experts. The unique aspect of ISWA-SWIS Winter School is that training is conducted both in-class and hands-on training over a two-week duration. The success and achievements of the 2016 ISWA-SWIS Winter School have been inspiring, motivating future possibilities and opportunities for the program. The SWIS team is committed to continuing the program every year, based on the success of the 1st winter school. Every year, participants are selected from developing countries that did not participate in previous years, as SWIS wants to reach out to more parts of the globe and involve more and more developing countries in sustainable waste management practices. In January 2020, SWIS completed its fifth successful SWIS' waste management training program. It has trained participants from more than 80 countries (including participants from every continent). In 2019, the Winter School program was awarded and honored as one of the best global sustainable waste management training programs by ISWA at the annual ISWA World Congress Conference during the Gala Dinner in Bilbao, Spain.

Finally, the proposed SMART facility will replace the existing waste management through open dumps or landfills as the sustainable resource management facility in every community in years to come. A SMART facility's success relies on conducting public outreach and education program, getting alignment with all stakeholders, politicians, and decision-makers, working with waste pickers, and creating jobs, and transferring of technologies needed at each stage.

MD Sahadat Hossain, Ph.D., P.E.

James Law, P.E., BCEE, IWM

Araya Asfaw, Ph.D.

REFERENCES

Hickman, M. (2018). Sweden Runs Out of Garbage, Forced to Import from Neighbors. Treehugger. https://www.treehugger.com/sweden-runs-out-of-garbage-forced-to-import-from-norway-4868335.

World Bank (2018a). *What a Waste: An Updated Look into the Future of Solid Waste Management*. https://www.worldbank.org/en/news/immersive-story/2018/09/20/what-a-waste-an-updated-look-into-the-future-of-solid-waste-management.

World Bank (2018b). *Population, total|Data*. https://data.worldbank.org/indicator/SP.POP.TOTL.

Zhu, D., Asnani, P.U., Zurbrugg, C. et al. (2007). *Improving Municipal Solid Waste Management in India: A Sourcebook for Policymakers and Practitioners*. The World Bank.

Series Preface

As the world's population, and with it the amount of resources we consume, continues to grow, it becomes ever more important to properly manage the "waste" that is generated by this growth. With the expanding volume and complexity of discarded domestic and industrial waste, and changing household consumption patterns, waste management is fast becoming one of the key challenges for the modern society.

According to some estimates, up to 2 billion people lack access to sound waste management. Uncollected municipal waste that ends up at illegal or improper dumpsites close to waterways and coasts generates marine litter, affecting marine ecosystems as well as the fishing and tourism sectors. Marine litter is primarily caused by the improper management of waste on land, which finds its way into the marine environment. Dumpsites, again caused by the improper collection and management of waste, can be sources of greenhouse gases and other short-lived climate pollutants. It is therefore recognized that the inadequate removal and treatment of waste poses multiple threats to human health and the environment. Particularly in low- and middle-income countries, open dumping and burning close to urban centers represent a substantial threat to human health and climate. However, advanced economies have shown that through effective waste management it is possible to significantly reduce these impacts and even, through recirculation of the materials in waste back into the production of new goods, minimize the broader impacts associated with consumption. Hence, the proper management of waste and resources is critical to successfully realizing several of the Sustainable Development Goals set out by the UN in 2015.

A sound and sustainable waste management system is a fine balance of a number of elements – technical, legislative, financial and business - carefully planned to unlock the economic potential of waste, including the creating of new jobs and development of new enterprises.

This series of books will address a range of topics that are integral to sound waste management systems. This will include technical solutions for waste collection and treatment, as well as addressing financing opportunities and the value of waste management, organizational and management challenges, and policy development and enforcement. The books will emphasize the need to adopt a holistic view of waste management by considering the total waste system, and then developing the most appropriate mix of infrastructure and services to manage the specific waste streams.

The Waste Crisis: Roadmap for Sustainable Waste Management in Developing Countries is the second book of the series and focuses mainly on technical aspects of sustainable waste management.

Björn Appelqvist
Chair of the ISWA Scientific and
Technical Committee (STC)

Any opinions expressed in this publication are solely those of the authors. They do not necessarily reflect the opinions or standpoints of ISWA, or its members, on any specific issue – unless explicitly stated.

Acknowledgments

The authors would like to take the opportunity to thank everyone who helped them implementing different waste management projects and who helped them enthusiastically to complete this book. Special thanks to:

- Brenda Haney, Current Solid Waste Director, City of Lubbock, Texas, USA, for her time and support reviewing the book manuscript extensively and providing valuable guideline to improve the quality of the book. Brenda is an active executive board member of SWIS (Solid Waste Institute for Sustainability). We also like to acknowledge her active role as Organizer and Presenter during ISWA-SWIS Winter School in all five years (2016–2020). Brenda has been playing an instrumental role for the success of the Winter School.

- Vance Kemler, Past Solid Waste Director, City of Denton, Texas, USA, for his support and help in implementing many sustainable waste management ideas in the city of Denton Landfill during his tenure as a director. Vance is an active executive board member of SWIS (Solid Waste Institute for Sustainability). We also like to acknowledge his active role as Organizer and Presenter during ISWA SWIS Winter School in all five years (2016–2020). Vance has been playing an instrumental role for the success of the Winter School.

- David Dugger (Past Landfill Manager, City of Denton, Texas, USA, and Current Manager of McKinney Landfill), Bill Sangster (Landfill Manager City of Irving, Texas, USA), Tiana Lightfoot Svendsen (Current Communication and Project Manager, US Plastics Pact, and past Recycling Director – City of Garland, Texas, USA), Dr. Patricia Redfearn (Solid Waste Director, City of Grand Prairie, Texas, USA) for their support and help implementing many sustainable waste management ideas in their cities. We also like to acknowledge their active participation and field demonstration during ISWA-SWIS Winter School in all five years (2016–2020).

- Our Winter School Ambassadors/participants from all over the world for sharing their waste management experience and stories from their countries selected from each Winter School, and their contribution to Chapter 3: Md. Shoriful Alam Mondal (Bangladesh), Thiago Villas Bôas Zanon (Brazil), Vishwas Vidyaranya (Colombia), Eshetu Assefa (Ethiopia), Medea Chachkhiani (Georgia), Visva Bharati Barua (India), Nour Kanso (Lebanon), Arely Areanely Cruz Salas (Mexico), Maria Ajmal (Pakistan), Soraia Taipa (Portugal), Dusan Milovanović (Serbia), Basem Abu Sneineh (UAE), and Tran Thi Diem Phuc (Vietnam).

- Tom Frankiewicz (US-EPA) and Silpa Kaza (World Bank) for their active help and support with international projects and their active participation as a speaker and panelist during ISWA-SWIS Winter School (2016–2020). Tom and Silpa are active executive board member of SWIS.

- Dr. Eshetu, Solid Waste Director of City of Addis Ababa for helping us with the tour of Reppie WTE Plant.

- Aditi Ramola, Technical Director of ISWA General Secretriat, Bjorn Appelqvist, Chair of Scientific Technical Committee (STC) of ISWA for their active help and support during the review process of the proposed book outline.

- Dr. Prabesh Bhandari (Geosyntec Consultants), Sachini Madanayake (Ph.D. student), and Muhasina Manjur Dola (Ph.D. student) for their help with different aspects of this book. We are very thankful to them for their hard work, dedication, and great service. Dr. Naima Rahman (SCS Engineer), Dr. Rakib Ahmed (ECS Consultants), Mumtahina Binte Latif (Atwell LLC), Sehneela Sara Aurpa (Ph.D. student) for their contribution and help.

- Ms. Ginny Bowers, our official reviewer/editor, for checking grammar and other aspects of the book.

- Sarah Higginbotham, Senior Commissioning Editor, John Wiley & Sons, Oxford, UK, for her keen interest in the book.

MD Sahadat Hossain, Ph.D., P.E.
H. James Law, P.E., BCEE, IWM
Araya Asfaw, Ph.D.

Chapter 1
Introduction

Sustainable solid waste management poses different challenges for developing and developed countries. In developing countries, 1/3 to 2/3 of the solid waste is currently dumped indiscriminately in streets and drains, where it can breed insects and rodents that transmit communicable diseases such as dysentery, typhoid fever, cholera, yellow fever, and plague (Zhu et al. 2007). Waste is traditionally burned in the streets and fields in and around city centers, producing dioxin emissions and damaging the air quality through soot emissions and the

The Waste Crisis: Roadmap for Sustainable Waste Management in Developing Countries,
First Edition. Sahadat Hossain, H. James Law and Araya Asfaw.
© 2022 John Wiley & Sons Ltd. Published 2022 by John Wiley & Sons Ltd.

climate through carbon dioxide emissions. Poorly managed waste has an enormous impact on health, the local and global environment, and the economy, and improperly managed waste usually results in downstream costs that are higher than the cost of managing the waste properly in the first place. Traditionally, as solid waste management practices improve, waste will begin to be placed in open dumps or landfills, which still poses human health/safety threats via disease vectors, water pollution, and explosive conditions. Uncontrolled landfills emit one-third of the anthropogenic methane emitted globally (ISWA Report 2021). Some of the current waste crises in developing countries are presented in the following section:

CRISIS 1

People have to cover their noses as they walk past the garbage that is piled up in and around city neighborhoods, residential areas, small businesses, and shopping complexes (Figure 1.1). In some places, dump trucks pick up the garbage once a

(a)   (b)

Figure 1.1 (a)/(b) Roadside uncollected waste becomes a nuisance for passerby
Source: (a) Fahim shaon/Wikipedia commons/Public Domain. (b) Courtesy of SWIS.

week, and in others, weeks go by without collecting it. An unfortunate but common practice is for people to burn their garbage in the corners of their neighborhoods and use it as a heating source during winter and/or just for fun. These practices can cause serious health and safety issues, and many times cause massive fires. Many airborne diseases are prevalent in these areas because of the open burning of garbage, but the residents are often unaware of them or their cause. Residents in the area are not happy about the practice, but the respective authorities are not taking any actions. It is essential that the practice of burning waste in open places as part of informal waste management system stops immediately.

<u>Questions</u>: (1) How can we stop this unhealthy and unsustainable social practice? (2) Why are the authorities not taking action?

CRISIS 2

Many metropolitan cities in South Asia, Africa, and Latin America have unusual flooding events (Figure 1.2). The floods are not during the regular flooding season, and in many cases, they occur in areas that are above the 100-year flood level. The roads and other infrastructures become completely flooded due to the heavy, continuous rainfall for several days and cause a nightmare for the city residents because of the (i) disruption of school/colleges and workplace schedules; (ii) difficulty in moving in and around the city by car, buses, or other road vehicles; (iii) serious health effects of highly polluted water (mix of rainwater and sewer water and many unhygienic floating materials on the water); and (iv) most seriously, the outbreak of many airborne and waterborne diseases immediately after the flood water recedes.

(a)

(b)

Figure 1.2 Flooding in cities disrupts life of residents. (a)/(b)Traffic flow through waterlogged roads.
Source: Palash Khan/The Daily Star. the city, Ronie/© Pixahive.com.

<u>Questions</u>: Why is this unusual flooding happening in the city? Can we avoid this situation?

CRISIS 3

Garbage disposal and management are the responsibilities of both regulators and private citizens, and the public's participation is vital to successfully resolve the problem. Indiscriminately throwing away trash without considering others or the environment creates problems regardless of the regulations that are in place (Figure 1.3).

(a)   (b)

Figure 1.3 (a) Trash adjacent to a garbage disposal container (Karthik 2018). *Source*: Ron Cogswell/Flickr. (b) Waste pickers collect trash from roadside. *Source*: Biswarup Ganguly/Wikipedia Commons/Public Domain.

CRISIS 4

A trash truck enters the dumpsite and hundreds of people (in this case, waste pickers) run toward or along the sides of the dump truck. This is an unbelievable but common daily view in many open dumps around the world and is comparable to hordes of people chasing a limo to get a glimpse of the celebrity inside. Ironically, the "celebrity" for the waste pickers is the trash that was thrown away by people for whom it had no value. Why are they running and possibly risking their lives? Because they want to retrieve as many recyclables or materials of value as possible. In some cases, the waste pickers jump on the dump truck (as presented in a photo from Tanzania, Figure 1.4), or even let the trash be dumped

Figure 1.4 Waste management hazards faced by waste pickers. Waste picker inside a collection truck in Tanzania.
Source: Courtesy of SWIS.

on them in order to get the most valuable items. Selling their collected trash in the market is their only means of making a living, and many of them are dying every day around the globe.

Question: How can we improve the waste pickers' deadly/unsustainable working conditions?

CRISIS 5

News Flash!! "Lebanese protest against waste-disposal crisis" Al Jazeera News, 26 July 2015.

The Naameh landfill, which served as the trash disposal facility for half of Beirut, Lebanon's population, was closed because it reached its capacity, and Sukleen, the main waste management company, stopped collecting rubbish because it had nowhere to dispose it. After that, garbage had piled up on every street corner in Beirut, and the stench of uncollected refuse was becoming unbearable (Figure 1.5a). In many cases, residents were burning trash on the street corners, which was unhealthy, as well as a serious safety concern (Figure 1.5b).

Figure 1.5 (a) Uncollected waste in the streets of Beirut.
Source: Dr. Amani Malaalouf, Winter School Participant from Lebanon. (b) Burning of trash during the Lebanon waste crisis. *Source*: Eliane Haykal/Adobe Stock.

Why were they in a situation like this? Primarily because the old dumpsites or landfills in many cities are reaching their maximum capacity, and the government or city officials are having difficulty finding a new location for a landfill. New landfill sites need to be identified by the city officials three to four years prior to the current site reaching its maximum capacity so that once the old dumpsite/landfill closes, the new landfill can begin accepting garbage. Finding a suitable location that is close to major metropolitan cities is becoming more challenging every year not only in Lebanon but also in many other developing/developed countries. *Space* is a big issue.

<u>Questions</u>: Why did not the authorities act ahead of time? How can we solve the major space problem?

CRISIS 6

"Death toll rises in Ethiopian trash dump landslide" – CNN, 15 March 2017
    In March 2017, Koshe, a very old dumpsite, collapsed in Addis Ababa, resulting in the deaths of more than 150 people who were living on or next to the dumpsite. The Koshe dumpsite started its operation outside the city almost 50 years ago; however, with time and expansion of the city's boundaries, the dumpsite was well within the city limits and became the home of many people who lived on or very close to the base of the dumpsite. The dumpsite reached

(a)   (b)

Figure 1.6 Landslide at the Koshe dumpsite in Addis Ababa. (a) Slope failure at Koshe. *Source*: Photo taken by author from site visit by SWIS. (b) Residential houses next to the failure site. *Source*: Photo taken by author from site visit by SWIS.

its maximum capacity and was closed in 2016, when the Sendafa Landfill, the first engineered sanitary landfill in Ethiopia, began operating just outside Addis Ababa. The new landfill ceased its operation after six months (July 2016) due to sociopolitical issues, however, and the city resumed dumping its trash at Koshe, which eventually caused a catastrophic failure and cost many lives (Figure 1.6a, b). Similar dumpsite failures cost lives in Sri Lanka, China, and Indonesia in recent years.

Questions: How can catastrophic failures such as the one that occurred at the Koshe dumpsite be avoided? What is the best way to close existing dumpsites without causing sociopolitical problems or creating a situation like the one in Lebanon in 2015? Can a waste management facility operate in one place perpetually for a long time?

CRISIS 7

A woman carries her little child on her back while she works at a dumpsite (Figure 1.7a). Many women and children like her work at dumpsites to collect bits and pieces from the pile of trash so that they can sell them in the market and make money for their families. Many of them also live on (Figure 1.7b) or next to the dumpsite, which is common and considered normal for people like her and millions of others around the world.

Figure 1.7 People living in dumpsites and making a livelihood from waste picking. (a) Children collecting waste. *Source*: Courtesy of SWIS, (b) young waste collectors taking a break while waiting for the next dump truck. *Source*: Toon van Dijk/Flickr.

<u>Questions</u>: Is it normal and health y for anyone to live on or next to a dumpsite, especially with kids? How can we improve their lives or livelihoods?

CRISIS 8

Many children are jumping into, swimming, and playing in a small pond (Figure 1.8a,b). They are having a great time during the summer heat where getting water to swim and play is not readily available to many in the world. Another group of kids is running around and playing next to the pond, and many of their friends are cheering them on. This seems like the perfect entertainment for the children, but the pond where they are playing is a leachate pond that is located at a dumpsite, where leachate and the surface runoff from rainwater are mixed. The playground is one corner of an open dump, and the children nor their parents have any idea about the serious health hazards that are present for those swimming in highly contaminated water and playing on and/or near an open dump.

Figure 1.8 Children near open dump of garbage: (a) people swimming in water polluted with trash. *Source*: Bibek2011/Wikipedia Commons/Public Domain, (b) girls playing with waste. *Source*: John Christian Fjellestad/Flickr. (c) A small boy playing in the landfill. *Source*: pxfuel. (d) A little boy looking towards the landfill. *Source*: https://www.pxfuel.com/en/free-photo-elam

<u>Questions</u>: Can we cast our cares aside and allow the practice to continue because they are not our kids, and does it not affect us when we are aware of the seriousness of the problem and the children's and parents' ignorance?

CRISIS 9

The news of China's restrictions on importing recyclables from foreign countries that became effective in January 2018 affected the waste recycling and waste management practices around the globe (Figure 1.9a).

Communities across the United States curtailed or halted their recycling programs, as the market for recycling products declined and the costs associated with recycling rose (Figure 1.9b). Many cities are sending their plastic waste and recyclables directly to the landfills, which are putting extra pressure on landfill operators as the increase in plastic waste, which consumes a higher volume of space in the landfill, will cause the landfill life to decrease significantly.

(a)   (b)

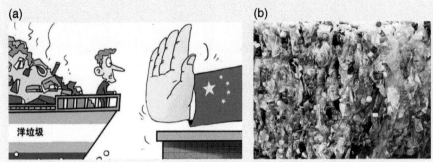

Figure 1.9 Effect of China ban. (a) China bans plastic waste.
Source: RTM World., https://www.rtmworld.com/news/china-to-ban-solid-waste-imports.
(b) Stacks of Plastic. *Source*: Hans/Pixabay.

<u>Questions</u>: How can we manage plastic waste sustainably? How can we create local markets for recyclables and avoid sending plastic waste to landfills so that we can benefit from a circular economy path?

All these waste crises are still happening in many developing countries around the world, and the questions raised are still unanswered. It is the intention of this book to raise awareness among the readers and to discuss some potential feasible and applicable solutions to these waste crises throughout the remaining chapters. And finally, for interested stakeholders and decision making around the world, answers to these questions are summarized in the last chapter of the book as a reference and starting point to solve their own issues and problems regarding better waste management practices for their own communities they live in.

In Developed Countries: Landfills typically occupy an area from several to hundreds of acres, and the current lack of available space for new landfills is a real problem for future waste management. Due to rapid growth and urbanization of cities beyond their current limits, many previously closed landfills, which were outside the city limit during their operation and closure, are now within the city limits. Opening a new landfill within the city limits often causes violent protests similar to those which occurred when demonstrations and protests against the opening of a new dump (landfill) on the slopes of Mount Vesuvius in October 2010 led to a riot and violent clash between the local residents and police.

Waste minimization and the reuse of existing landfills are keys for sustainable urban development, but even the most preferred choices of waste management, recycling and reuse, have inherent problems. For instance, Sweden embraces recycling; however, the incineration of solid waste provides power to 250 000 of their homes and heats 810 000. Recycling is so effective in Sweden that only 4% of all waste generated in the country is landfilled. However, the Swedes ran out of garbage needed for the incinerators and have had to import it from Norway (Hickman 2018).

Waste-to-energy (WTE) or incineration can provide a lucrative solution to waste management as WTE plants can address the issue of land/space in both developed and developing countries. WTE or incineration systems require less land or space than any other available system and can process a large volume of waste in a single processing plant. It is an appropriate technology for waste management in developed countries such as Europe, the United States, Japan, and South Korea, where the waste is relatively dry and the substantial presence of plastics, paper, and wood makes it a good source of combustible materials. The waste in developing countries has a higher percent of food waste (more than 70%) and corresponding moisture content than that of the waste in European countries, however, so generating power through WTE or incineration, using this low calorific (organic waste) and highly wet waste may not be applicable or cost-effective.

The hierarchy of waste management may not be the same for all countries or all kinds of solid waste. In other words, *one solution does not fit all.* The handling of solid waste requires solutions that are flexible, but robust enough for urban sustainability.

Regions of Africa, Latin America, South Asia, East Asia, the Pacific Islands, and the Caribbean are urgently in need of help in mitigating problems associated with increasing population growth, urban consumption, and waste production, and if safe waste management practices are not incorporated into their development plans, the proliferation of poor health and sanitation conditions will persist.

A clear understanding of waste generation, collection, and management practices, as well as a roadmap for developing a sustainable waste management plan in developing countries, is vital for environmental sustainability, good health, the safety of waste workers and people living in communities near dumpsites, and above all, for creating healthy urban cities across the globe. And this book provides a roadmap for sustainable waste management for developing countries.

REFERENCES

ISWA. (2021). A Roadmap for closing Waste Dumpsites The World's most Polluted Places, December 12. https://www.iswa.org/closing-the-worlds-biggest-dumpsites-task-force/?v=7516fd43adaa. (accessed 7 February 2022).

Hickman, M. (2018). Sweden runs out of garbage, forced to import from neighbors. *Treehugger*. https://www.treehugger.com/sweden-runs-out-of-garbage-forced-to-import-from-norway-4868335.

Karthik, A. (2018). Remains of the day: diwali aftermath. *Deccan Chronicle* (8 November). https://www.deccanchronicle.com/nation/current-affairs/081118/remains-of-the-day-diwali-aftermath.html (accessed 7 February 2022).

Zhu, D., Asnani, P.U., Zurbrugg, C. et al. (2007). *Improving Municipal Solid Waste Management in India: A Sourcebook for Policymakers and Practitioners*. The World Bank.

Chapter 2

Current Waste Management Practices

2.1 URBANIZATION AND WASTE GENERATION

The world is rapidly moving toward urbanization, and the amount of municipal solid waste (MSW), one of the most important by-products of an urban lifestyle, is growing even faster. Based on the latest data available, global waste generation in 2016 was estimated to have reached 2.01 billion tons (4.4 trillion lb) (World Bank Report 2018a). By 2030, the world is expected to generate 2.59 billion tons (5.71 trillion lb) of waste annually, and by 2050, waste generation across the world is expected to reach 3.40 billion tons (7.5 trillion lb).

The United Nations (UN 2018) defined the process of urbanization or "urban transition" as the shift in population densities from a rural/agriculture-based economy to a denser population with an industrial and service-based economy. Urbanization is the result of an increase in population and a shift from agricultural employment to jobs focused on industry/manufacturing, migration, and immigration (Vij 2012). The urban population has increased steadily over the last decade. According to an urbanization

The Waste Crisis: Roadmap for Sustainable Waste Management in Developing Countries, First Edition. Sahadat Hossain, H. James Law and Araya Asfaw.
© 2022 John Wiley & Sons Ltd. Published 2022 by John Wiley & Sons Ltd.

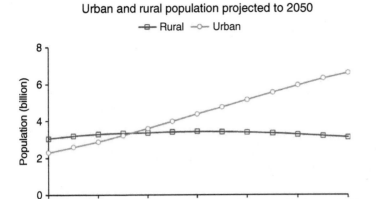

Figure 2.1 Projected global urban population (Reproduced using UN World Urbanization Prospects 2018).

report by the UN (2016), the world's urban population has been greater than the rural population since 2006 (Figure 2.1). The projected urban population in the world is 6.6 billion by 2050, which represents an increase of more than 50% over the next 30 years.

There are striking differences in urbanization pattern between the more developed regions and the less developed regions (UN 2018). The urban population of the less developed regions has been growing considerably faster than that of the more developed regions as presented in Figure 2.2. In 2018, three times as many urban dwellers were estimated to live in the less developed regions as in the more developed regions (3.2 billion versus 1.0 billion). As the developing world becomes increasingly urbanized, by 2050, with 5.6 billion urban dwellers, the less developed regions are projected to have 83 percent of the world's urban population and 87% of the total world population. Figure 2.3 clearly shows that the urban population change is more significant in Asia, sub-Saharan Africa, and Latin American countries. More large cities (population more than 1 million) and Megacities (population more than 10 million) are in these regions.

Considering the change in urban population, we will focus on specific cities from Asia, Sub Saharan Africa, and Latin America. The %age of the urban population in various selected regions is presented in Figure 2.4. The rate of increase in the urban population is significantly higher in developing countries than in developed countries. The projected absolute increase in the United States is 14% (between 1990 and 2050), whereas the projected change in urban population in China is 54% and Ethiopia is 26%, respectively, between 1990 and 2050 (Ritchie 2018).

Urbanization is also a result of migration (IOM 2015). Figure 2.5 shows the changes in the urban and rural populations over a time period of 60 years. The developing countries show a projected increase of more than 100% in their urban population

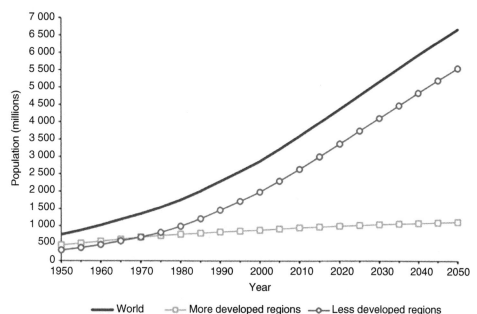

Figure 2.2 Estimated and projected urban populations of the world, the more developed and less developed regions (UN World Urbanization Prospects 2018).

from 1990 to 2050. The reduction in employment opportunities in rural areas and the emerging industries and services in urban areas have fueled the shift of the population from rural to urban, and the migration from rural to urban areas changed the employment opportunities from being agriculturally based to being related to industry, manufacturing, and services. The decrease in agricultural employment has been more prominent in developing countries than in developed countries. A 30% reduction in agricultural employment was observed in Bangladesh, which had a 15% increase in its urban population over a span of 26 years (Figures 2.6 and 2.7). The projected relative change in urban population in Bangladesh, however, was 200% (Figure 2.5).

Urbanization has generally been a positive force for economic growth, poverty reduction, and human development. Diverse and well-educated workforce and high concentration of business make cities perfect place for young entrepreneurs to thrive. With higher%ages of young urban dwellers, economic activity increases significantly in urban areas, which contributes to gross domestic product (GDP) growth of urban population in a country as presented in Figure 2.8. Increase in population in urban areas and GDP growth result in increase in consumption ultimately leading to an increase in waste generation. Figure 2.9 shows the relationship between GDP per capita and household consumption for few selected countries. As expected, household consumption is increasing with GDP.

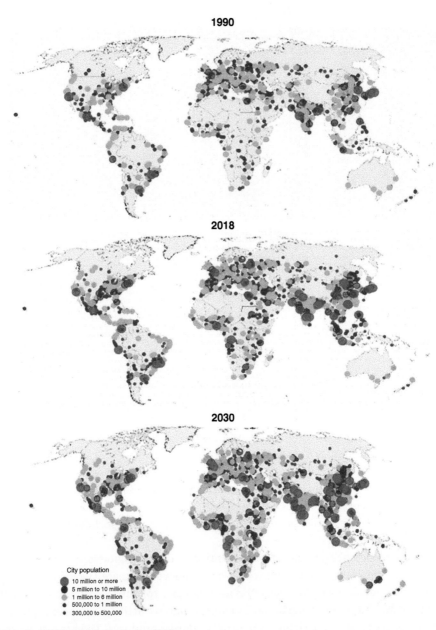

1990

2018

2030

City population

- ● 10 million or more
- ● 5 million to 10 million
- ● 1 million to 6 million
- ● 500,000 to 1 million
- ● 300,000 to 500,000

* For cities with 300 000 inhabitants or more 2018.

Figure 2.3 Global perspective on distribution of urban settlement (UN 2018; https://population.un.org/wup/Maps).

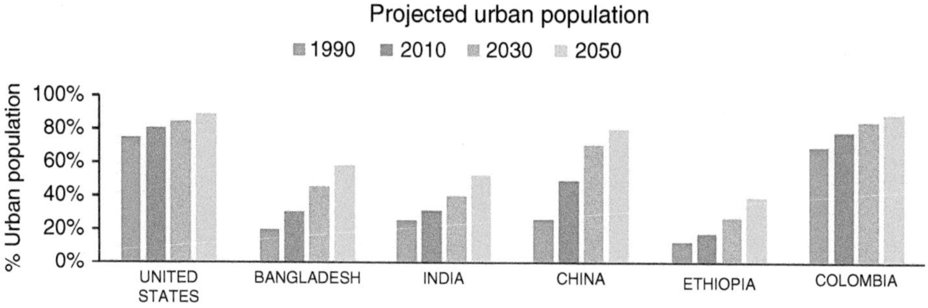

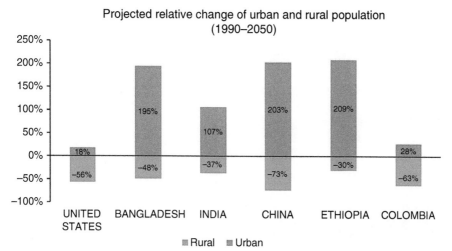

Figure 2.4 Projected urban population in cities.
Source: Data from UN World Urbanization Prospects 2018.

Figure 2.5 Relative change of urban and rural population (1990–2050).
Source: Data from UN World Urbanization Prospects 2018.

Economies of scale in urban areas and technological innovation have facilitated the development of infrastructure such as roads, piped water, and electricity, as well as basic services such as education and healthcare, all of which are essential to achieve the sustainable development goals. However, waste collection, processing, and disposal, which are of great importance for development of safe, healthy, and smart urban cities, have remained almost similar or got worse in many developing countries in past 30–40 years. In some countries, major cities are dangerously moving toward a major waste crisis. For a sustainable urban future with less carbon emission and clean air quality, sustainable waste management needs to be a major priority for many countries around the globe.

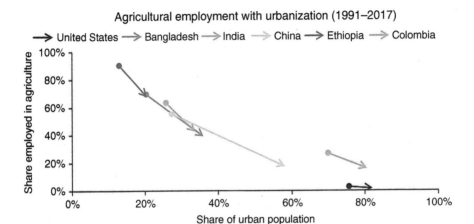

Figure 2.6 Change of agricultural employment with urbanization.
Source: Data from World Bank – World development indicators 2021.

(a)   (b)

Figure 2.7 (a) Agriculture-based (https://www.worldbank.org/en/results/2016/10/07/bangladesh-growing-economy-through-advances-in-agriculture). *Source*: Dr. Matthias Ripp/Flickr. and (b) manufacturing-based employment in Bangladesh (https://www.ifc.org/wps/wcm/connect/news_ext_content/ifc_external_corporate_site/news+and+events/news/insights/bangladesh-garment-industry). *Source*: Mohammad Zahin Rahman/IFC.

Based on the urbanization pattern presented in previous sections, it is not surprising to see that the projected increase in waste generation is highest in low-income countries (more than triple by 2050), as they are expected to have higher economic and population growth (World Bank Report 2018b) as presented in Figure 2.10. More cities in low-income countries are experiencing extensive urbanization, as discussed in the previous section, and the combined effects of urbanization, GDP growth, purchasing power, and waste generation have increased significantly.

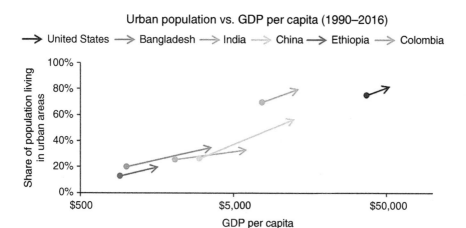

Figure 2.8 Relationship between urban population and GDP per capita:
Source: Data from UN World Urbanization Prospects 2018.

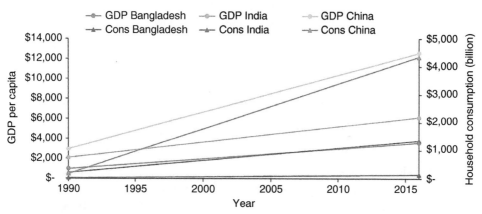

Figure 2.9 Relationship between GDP per capita and household consumption (World Bank data; https://www.theglobaleconomy.com/rankings/household_consumption_dollars).
Source: Data from TheGlobalEconomy.com [24].

Upper-income and high-income urbanization peaked in developed countries during the same time period, and economic growth continued, although not as rapidly as in the developing world. Recycling, reuse, source depletion, and diversion of solid waste from final disposal are more efficient in high-income countries because most of the citizens are cognizant of waste management issues and are serious about protecting the health of their communities and the environment. The conclusion is that waste generation is expected to be the lowest in wealthy countries (less than 1.3 times by 2050), as they have already reached the peak of their economic activity. Moreover, population growth is projected to be less; in some cases, they are experiencing negative population growth.

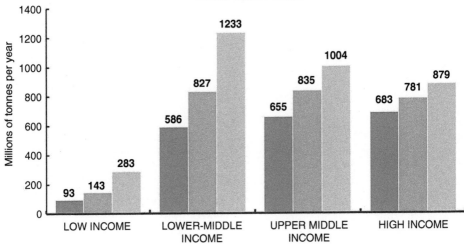

Figure 2.10 Urban waste generation by income level (Redrawn using data from World Bank 2018a).

Note: 1 million ton = 2.2 billion lb.

In terms of region, the sub-Saharan African, South Asian, Eastern European, and some Latin American countries are low-income and lower middle-income countries where urbanization is happening at a much faster rate than in any other parts of the world (Figure 2.3). There will be more new megacities (with populations greater than 10 million) and more business activities, as well as a larger increase in the GDP growth. Local and international communities are moving their businesses to megacities, and the population is migrating at a fast pace. An associated increase in waste generation is expected to occur with the increase in population and purchasing power (Figure 2.9), as is clearly reflected in World Bank Report (2018a). The sub-Saharan Africa and South Asia regions are projected to see an increase in waste generation levels that are approximately triple and double, respectively, by 2050, with higher economic activity and urbanized population growth (World Bank Report 2018b). Regions with higher income countries (North America, Europe, and Central Asia) are expected to see a much slower rate of increase in waste generation levels (Figure 2.11).

2.2 WASTE COLLECTION

With the exponential growth of population in the new megacities and urban cities, the counties/cities are under tremendous pressure to meet the additional demands. Housing construction, electricity and power generation, roads and highways, water

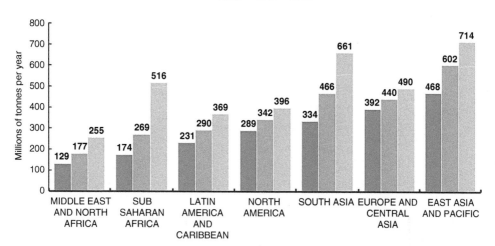

Figure 2.11 Waste generation by region (Redrawn using data from World Bank 2018a).

Note: 1 million ton = 2.2 billion lb.

infrastructure, and transportation vehicles are receiving high priority, and all of these growth-related activities are generating more waste and creating potentially unhealthy and unsafe environments. Waste management is a low priority for most of the new urban cities and that causes serious challenges for sustainable urban development. Moreover, with their newfound wealth and prosperity, the citizens are focusing more heavily on consumption rather than on environmental health and safety. The general mindset of "it is the government's responsibility to clean our mess" is not helping either, and along with the combined effects of increased waste generation, the ignorance of the city officials, and a minimum investment in collection vehicles lead to lower collection of waste. The World Bank report points toward similar findings.

Low-income countries have low collection rates (around 39%), while high-income countries have higher collection rates averaging 98% (Figure 2.12). At present, most of the developed world (Europe, Latin America, the Caribbean, North Africa, and Central Asia) has a collection rate of more than 70%, and South Asia and sub-Saharan Africa have a collection rate of less than 50% (Figure 2.13). Ironically, the regions where collection rates are the lowest are expected to experience the highest increase in waste generation (triple in sub-Saharan Africa and double in South Asia).

Figure 2.14 clearly demonstrates the impending waste crisis and its serious impact on human health, the environment, and most importantly, sustainable urban development.

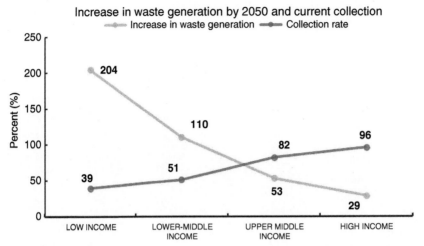

Figure 2.12 Average waste collection by income level (Data from The World Bank 2018a).

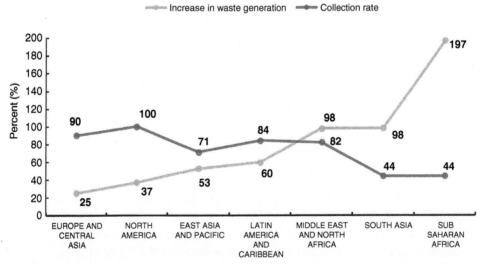

Figure 2.13 Average waste collection by region (Data from The World Bank 2018a).

Based on the World Bank report and the review of other related reports, waste collection, which is a necessary service that is expected to be provided by cities, countries, or regional authorities, is the lowest in sub-Saharan Africa and South Asia (Figure 2.14). The ISWA report (2016) also pointed out waste collection is a major crisis in these countries and unless an effective and sustainable plan is developed and implemented, a healthy urban city is only a dream.

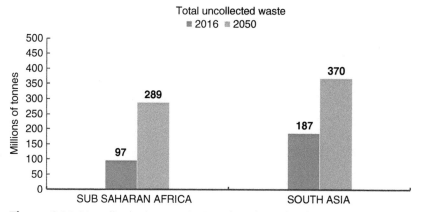

Figure 2.14 Uncollected waste in South Asia and sub-Saharan Africa (Data from The World Bank 2018a).

Note: 1 million ton = 2.2 billion lb.

Uncollected solid waste has a devastating effect on the environment, economy, and health of the most distressed sectors of the population. Improving waste collection methods and better managing the collected waste are critical steps to creating healthy and clean urban cities globally and will eventually result in creating jobs and alleviating poverty.

2.2.1 Why Waste Collection is Low in Developing Countries

Many factors can affect the waste collection rate: the income level of a region, the education level of the residents, the commitment to environmental health and safety, the availability of waste collection vehicles, trained workforce, social programs for creating awareness among the residents, etc. This section focuses primarily on the waste collection practices in developing countries.

2.2.1.1 Waste Collection Flow

Many factors contribute to waste collection being the lowest in developing countries. First, let us review the waste-collection-to-disposal flow to identify the potential problems of each component and how it affects the collection of waste from house to truck. Waste can be collected in many different ways: (i) house-to-house, (ii) community bins, (iii) curbside pickup, (iv) self-delivered, and (v) contracted and delegated service (World Bank 2012). Options 1, 2, and 4 are predominant in developing countries, and combinations of them are used by many communities. On the other hand, options 3 and 5 are primarily used by developed countries. The household waste collection is presented here for both developed and developing countries.

In developed countries, waste is placed outside houses or business developments and is collected by trained workers who drive/accompany waste collection vehicles that collect the garbage from community bins. The residents are responsible for placing the garbage in bins and moving them curbside, outside their homes/businesses. The trained city workers run from house to house, collecting the trash and putting it in the collection vehicle until all of the community's trash has been picked up. It is usually only one step from the bin to the truck. There are a variety of collection vehicle types that can be used for these purposes – (a) rear loader, (b) automated side loaders, or front loaders – that are primarily used for business or commercial collection (Figure 2.15).

The waste collection flowchart for developed countries is shown next (Figure 2.16).

In developing countries (low- and/or low-middle-income countries), waste is collected by community waste workers who go from house to house and place it in small vans.

(a) (b)

(c)

Figure 2.15 (a) Rear loader garbage truck (www.heil.com). *Source:* Jason Lawrence/Flickr. (b) Automated side loader garbage truck (https://www.wikiwand.com/en/Garbage_truck). *Source:* carol/Wikimedia Commons/Public Domain. (c) Front-end loader garbage truck (www.bakerswaste.co.uk/wp-content/uploads/2016/01/FEL_Northampton_OldSite.jpg). *Source:* Bakers Waste Services.

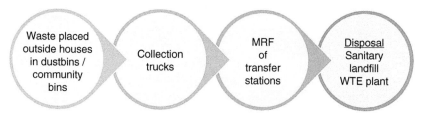

Figure 2.16 Waste collection flowchart for developed countries.

Figure 2.17 Use of manual labor for trash collection in Bangladesh (Photo taken by SWIS member).
Source: Courtesy of SWIS.

Once the van is full (which happens quickly), they take the waste to nearby community bins. The city-operated collection trucks use manual labor, as seen in Figure 2.17, to collect all the trash and place it in the trucks, where it is compacted manually resulting in low-compaction levels. The three-step method utilized from households to collection trucks is very time consuming. More waste and a less efficient system are the first major reasons for the low rate of waste collection in developing countries. Waste collection in developed countries is presented in Figure 2.18.

Waste collection flowchart for developing countries is presented in Figure 2.19.

2.2.1.2 Waste Collection Vehicles and Their Capacity

The automated collection vehicles (Table 2.1) have a high compaction capacity and can carry up to 20–30 tons (44 000–66 000 lb) of garbage, whereas the manually compacted collection trucks can carry 2–3 tons (4400–6600 lb) of garbage. Therefore,

Figure 2.18 Waste collection in developing countries. (a) Waste collection workers loading waste into trucks (http://percolado.blogspot.com/2013/09/dhaka-banglagesh.html). *Source*: Courtesy of SWIS, (b) manual waste collection vans. (https://vrindavanactnow.org/religious-leaders-act-now). *Source*: Courtesy of SWIS, (c) manual trucks for waste collection (https://commons.wikimedia.org/wiki/File:Kathmandu-M%C3%BCllabfuhr.jpg). *Source*: Wikimedia Commons.

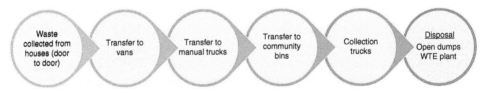

Figure 2.19 Waste collection flowchart for developing countries.

one truck load of garbage in developed countries is equivalent to ten truckloads of garbage in developing countries (Figure 2.20). This also means that the manual trucks need ten times more trip time, which includes more time on the road and increased traffic along roads and highways for collecting an equivalent amount of garbage. Without ten times more trucks and trained personnel or automated collection trucks,

Table 2.1 Features of different collection vehicles.

Collection vehicle	Capacity	Compaction level	Used locations	Advantages	Disadvantages
Van	1.42–1.7 m³ (50–60 cu-ft)	Very low	Developing countries	Low maintenance cost, easy operation	Low capacity, manual operation, less efficiency, high collection time, hazardous to health, less hauling capacity
Medium size truck	2.27–4.81 m³ (80–170 cu-ft)	Low	Developing countries	Low initial cost, easy maintenance	Low compaction, less efficiency, not environment friendly
Truck	4.25–10.76 m³ (150–380 cu-ft)	Low	Developing/developed countries	Easy maintenance	Manual collection, high collection time, fewer safety measures
Advanced Garbage Truck	9.8–19.6 m³ (350–700 cu-ft)	High compacting ratio - 6:1	Developed countries	High capacity, less collection time, high safety measures	High initial investment High maintenance cost
Fully Automated Waste Collection Truck	9.8–19.6 m³ (350–700 cu-ft)	High, 80–90 lb per cu-ft	Developed countries	Fully automated, high efficiency, less collection time, very high safety measures	High initial investment, very high maintenance cost, requires high skilled operation

Note: 1 cu-ft = 0.028 m³ / 1 lb = 0.45 kg.

Figure 2.20 Types of waste collection vehicles: (a) rickshaw van (https://water1st.org). *Source*: Water 1st International, (b) manual garbage truck (www.dailymail.co.uk). *Source*: Kelly L/Pexels, (c) front-loader garbage truck (https://tractors.fandom.com/wiki/Waste_collection_vehicle). *Source*: Tractor & Construction Plant Wiki, (d) rear-loader garbage truck (www.heil.com). *Source*: Heil – An Environmental Solutions Group Company, (e) automated side-loader garbage truck (https://thereaderwiki.com/en/Waste_collection_vehicle). *Source*: carol / Wikimedia Commons / Public Domain, (f) grapple truck (https://mvccolombia.co). *Source*: MY COLUMBIA, (g) pneumatic collection truck (https://www.wikiwand.com/en/Garbage_truck). *Source*: G®iffen / Wikimedia Commons / Public Domain, (h) roll-off truck (https://en.wikipedia.org/wiki/Roll-off_(dumpster)). *Source*: Boomer77 / Wikimedia Commons / Public Domain, and (i) gully emptier (https://ndbrown.co.uk). *Source*: Anthony Appleyard/Wikipedia Commons / Public Domain.

(g) (h)

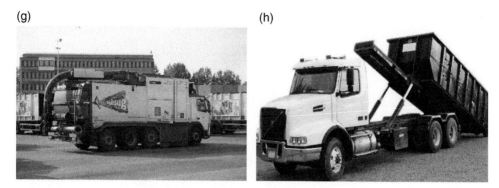

(i)

Figure 2.20 (Continued)

waste collection will remain at its current low rate. Therefore, the lack of advanced waste collection vehicles is the second main reason that waste collection is low in developing countries.

2.2.1.3 Traffic Situation in Developing Countries

Once the garbage truck is fully loaded, it proceeds to the next stop, which varies depending on the design of the waste management system (transfer stations/Material Recovery Facility [MRF], processing facilities, or final disposal). The amount of traffic encountered along the route can be a major factor in developing countries; in developed countries, landfills/WTE/MRF facilities are located in areas that are easily accessible and avoid traffic congestion (Figure 2.21).

Figure 2.21 (a) Typical highway traffic condition in Texas. *Source*: Courtesy of SWIS. (b) Route taken by a waste disposal truck near the Hunter Ferrell Landfill in Irving, Texas. *Source*: Courtesy of SWIS. (c) Waste disposal truck entering the landfill (Photos taken by SWIS Member). *Source*: Courtesy of SWIS.

Traffic routes are usually planned to avoid or minimize traffic congestion; however, in developing countries, especially new urban centers or megacities where the population growth has outpaced the city's development, traffic congestion is very serious (Figure 2.22). The additional trucks that are needed to increase the collection rate exacerbate the normal traffic conditions; therefore, city officials make their trash collection a lower priority to avoid traffic congestion during peak times, which may hinder the efficiency of the trash collection. As a result, traffic congestion in new urban city centers or megacities is another major reason for the lower waste collection rate in developing countries.

2.2.1.4 Trained Waste Collection Workers

In developing countries, the community level waste collectors of household trash are not trained for their jobs and their routes are not planned for efficient collection.

Figure 2.22 Typical traffic congestion in megacities located in developing countries (https://www.thedailystar.net/city/dhaka-traffic-jam-congestion-eats-32-million-working-hours-everyday-world-bank-1435630). *Source*: mohd firdaus zulkefili/ EyeEm/Adobe Stock.

The next tier waste collectors use vans to take the trash from community trash bins or locations. Both the collection workers lack the training to efficiently collect, compact, and manage the waste collection procedure. Moreover, once the garbage is finally dumped into the working face, lack of proper training hinders the operation in the working face. This can cause traffic jams at open dumpsites and high turnaround times for garbage collection trucks that result in decreased waste collection rates and efficiencies.

2.2.1.5 Lack of Social Awareness and Illegal Dumping

A critical component of any waste management program is public awareness and participation. In developing economies, the residents are usually unaware of the need for sustainable waste management. Most of them have the general attitude that when you throw something away, someone else will take care of it. With this mindset, no one is willing to follow the regulations set by the city, state, or other government officials, and waste is visible everywhere. Illegal open dumping and open burning of waste are the main practices that are implemented for waste treatment and disposal in developing countries (Ferranato and Torrette 2019), and such unsustainable practices include every waste fraction, such as MSW, hazardous waste (HW), construction & demolition (C&D) waste, used tires, used batteries, and industrial waste. Each spreads a specific contaminant concentration in the soil, water, and air environments and increases health and occupational risks. In summary, creating an awareness among the public of the vital need for their participation is essential for increased waste collection, sustainable waste management, and healthy urban cities.

2.2.1.6 Absence of Regulations and/or Lack of Interest in Implementing Them

In many cities and countries, there are virtually no regulations that control waste collection and disposal for either residential or commercial MSW. Garbage is rarely even collected on a regular basis. Regulations vary from country to country and from town to town, but even when there are minimal regulations in place, there is no intent for the city officials to enforce or implement them and they openly ignore open burning. Based on an informal discussion, it appears that some of the reasons for their actions are that (i) the open burning reduces the volume of waste and therefore reduces the city's burden for managing it, (ii) the reduced volume of waste reduces the drain on their already limited budget, and (iii) they do not have the workforce to enforce controlling or reducing open burning. A small bribe from an illegal trash dumper or illegal open burner often trumps enforcement of official regulations anyway. Frequently, a lack of funds prevents municipalities in such countries from ever being able to even create a proper waste management system. To conclude, the main issues that surround the inadequate development and enforcement of regulations and insufficient funding stem from the ignorance, or in some cases deliberate ignorance, of the city officials about:

1. The harmful effects of illegal burning on the people living nearby, which in many cases are the most economically distressed population of the city.

2. The need for proper waste management and creating cleaner and more sustainable urban and rural cities.

3. Every citizen's deserved opportunity for a better life regardless of their current economic condition.

4. The lack of status and poor salaries associated with the profession that discourages qualified employees who have the ability or the training to manage an effective system.

The ignorance of city officials and regulators is one of the main reasons for low waste collection rates. A lack of available funding is often presented as the main problem; however, it seems likely that the lack of understanding of the need for sustainable waste management drives the officials to allocate an inadequate budget for this sector.

2.2.2 Consequences of Having Lower Waste Collection and Associated Open Dumping

Low waste collection translates to uncollected trash being strewn around the city and causing serious health, environmental, and economic issues. The waste finds its way into local waterways, roads, and drainage systems that become breeding grounds for disease. Some of the major problems caused by uncollected waste are presented in the following.

2.2.2.1 Polluted Water Channels/Lakes/Rivers/Oceans

Uncollected garbage is finding its way into lakes, rivers, and other bodies of water that become breeding grounds for mosquitos or sources of other infectious diseases when they become clogged with garbage (Figure 2.23). The situation is an alarming one in the emerging African and Asian megacities, where hundreds of millions of people are expected to move within the next 25 years, reshaping what is called urbanization and creating new challenges for civilization (ISWA 2012). People living in these areas are unaware of the serious health hazards, and they are spending a significant portion of their income on associated medical expenses.

Figure 2.23 (a) Polluted rivers in developing nations. *Source*: McKay Savage/ Wikipedia Commons/Public Domain. (b) Polluted riverbanks. *Source*: Shubert Ciencia/ Wikipedia Commons/Public Domain. (c) People riding on boats through polluted water bodies. *Source*: The Innovation Diaries.

2.2.2.2 Flash Flooding During Rainy Seasons

In recent years, there have been unusual flash flooding events in many metropolitan cities in South Asia, Africa, and Latin America. The floods did not occur during the regular flooding season, and in many cases were in cities that are above the 100-year flood level. The heavy rainfall that continued for several days caused roads and other infrastructure to flood and created a nightmare for the residents by disrupting the schedules of their schools, colleges, and workplaces. Additionally, it was difficult or impossible to move around the city by car, buses, or other road vehicles. Most significantly, the flooding created highly polluted water (mix of rainwater and sewer water and many unhygienic floating materials on the water), which initiated the outbreak of many airborne and waterborne diseases immediately after the flood waters receded. Dhaka, the capital of Bangladesh, has a population of approximately 20 million, making it one of the world's most populous cities (World Population Review 2016); however, it, like other megacities in Asia and Africa, faces major challenges on its path to development due to environmental pollution caused by the lack of a sustainable waste management system. Dhaka has made great progress on many fronts, including (i) massive construction of apartment buildings to house more people, (ii) an increase in the number of academic institutions within the city (Figure 2.24a), (iii) an increase in outdoor dining/entertainment facilities (Figure 2.24b), (iv) an exponential increase in the number of shopping centers (Figure 2.24c), (v) an increase in the GDP and the people's purchasing power, and (vi) an unbelievable increase in the number of private cars (Figure 2.24d). All of these also unfortunately give rise to an exponential increase in consumption and subsequently waste generation.

Although, it recently adopted a solid waste master plan, most areas of the city lack sufficient waste collection services. Only 40–60% of Dhaka's generated solid waste (around 10 000ton per day/20 million lb per day) is collected and transported to one of the city's two dumpsites or limited sanitary landfills. Approximately 5000tons (10 million lb) per day of uncollected waste is dumped in open places, on roads and streets, in bodies of water, and in most public areas, creating public health and environmental hazards for people living in the city and hampering the city's esthetic appearance (Figure 2.25).

Congestion of the limited road networks and the lack of good collection equipment are causing Dhaka city to have a very low rate of waste collection. Uncollected waste is getting into city's drainage system, and plastic bags are clogging the city's drainage system and causing flash flooding. This is hindering the city's objective to become a world class urban city and can also deter the interest of foreign investors and creates a bad image for the country as a whole. An increase in the amount of waste that is collected and sustainable waste management are therefore vital for the city's future development.

(a) (b) (c) (d)

Figure 2.24 (a) University of Asia Pacific campus in the city (https://commons. wikimedia.org/wiki/File:North_South_University_-_panoramio_-_mohigan.jpg). *Source*: Wikimedia Commons. (b) Shopping centers and roadside restaurants that are becoming a common sight in Dhaka city (https://lifeofmudita.com/travel-dhaka). *Source*: joiseyshowaa/Wikipedia Commons/Public Domain. (c) Large shopping malls (Bashundhara City) in the heart of the city (https://dailyasianage.com/news/39872/the-second-largest-shopping-mall-in-bangladesh). *Source*: Ragib/ Wikipedia Commons/Public Domain. (d) Typical traffic conditions during the peak hours (https://www.thedailystar.net/city/dhaka-traffic-jam-alert-public-suffering-2-and-half-hours-and-still-the-road-1437955). *Source*: Ragib Hasan/Wikipedia Commons/Public Domain.

2.2.2.3 Serious Health Hazards

Poorly managed waste has an enormous impact on the health of the community, the local and global environment, and the economy, and often results in downstream costs that are higher than they would have been had the waste been managed properly in the first place. Unregulated or poorly managed waste is burned openly in the

(a) (b)

Figure 2.25 Flooding in cities disrupts life of residents. (a)/(b)Traffic flow through waterlogged roads. *Source*: Palash Khan/The Daily Star. the city, Ronie/© Pixahive.com.

streets and fields of many cities and in and around city centers, producing dioxin emissions and damaging the air quality through soot emissions and the climate through carbon dioxide emissions. Uncontrolled waste and waste dumpsites emit one-third of the anthropogenic methane emissions globally (ISWA 2012). Rotting or untreated waste exacerbates the spread of malaria, dengue fever, cholera, typhoid, respiratory, and skin diseases, while dioxins may insidiously cause cancer wherever they are deposited (Figure 2.26).

2.3 PROCESSING AND FINAL DISPOSAL

The options for processing and final disposal of MSWs across the globe include landfilling, recycling, waste-to-energy, incineration, dumpsites, composting, and others. The waste management hierarchy is presented in Figure 2.27, and the total global waste disposal rates are presented in Figures 2.28 and 2.29 (World Bank 2018a).

Even though landfilling or a dumpsite is the least preferred choice for final disposal of waste, approximately 50% of the waste generated is disposed of in landfills in high-income countries, and the remaining 50% is diverted from landfills by composting, recycling, waste-to-energy, etc., as presented in Figure 2.29. In middle- and lower-middle-income countries, 90% of the waste goes to dumpsite or landfills and only 10% is diverted from them. Globally, 70% of all waste is dumped into open dumpsites or landfills. The global waste processing and management practices can be categorized as follows:

1. Sanitary landfilling (50%) – mainly in developing countries. In Europe, more than 70% countries send more than 50% of their waste to landfills (Figure 2.30).

2. Open dumps and landfills (90%) – mainly in developing and upper-middle-income countries.

3. Recycling (up to 30%, with the exception of a few European countries that recycle more than 90% of their waste).

Figure 2.26 Uncollected waste on the streets in Bangladesh. (a) Uncollected waste near residential buildings (https://www.quora.com/Why-are-cities-in-developing-countries-so-dirty-What-does-it-take-to-keep-a-city-clean). *Source*: Joegoauk Goa/Flickr. (b) Odors emanating from open dumps that are a nuisance for pedestrians. *Source*: Eyal Ofer/Flickr. (c) Uncollected waste outside garbage cans (Hossain 2009). *Source*: EUvin/Wikipedia Commons/Public Domain.

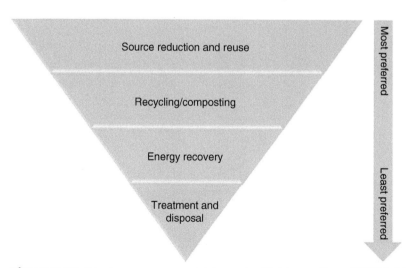

Figure 2.27 Waste management hierarchy (Redrawn from USEPA 2013).

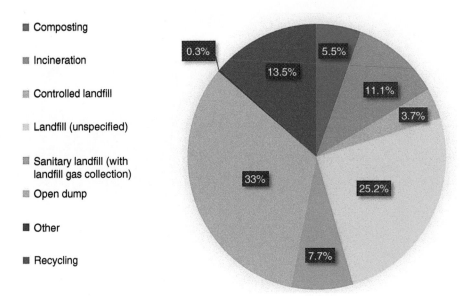

Figure 2.28 Global waste management practices (Redrawn from World Bank 2018a).

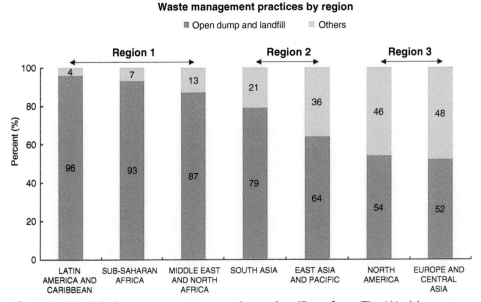

Figure 2.29 Global waste management by region (Data from The World Bank 2018a).

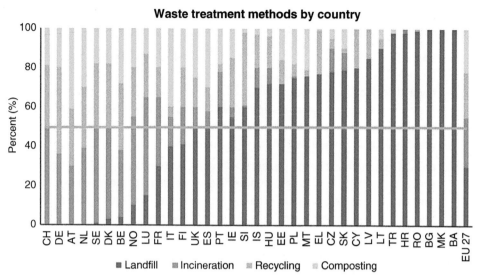

Figure 2.30 Processing of household waste in Europe (EU-27), 2009 (Redrawn from Eurostat 2009).

4. Waste-to-energy (developed and developing countries) and incineration.

5. Composting.

Landfilling is still the predominant mode of waste management and processing; however, it too has some major problems, as presented in the following.

2.3.1 Problems with Landfilling in Both Developed and Developing Countries

According to the US Environmental Protection Agency (EPA), US residents, businesses, and institutions produced more than 250 million tons (550 billion lb) of MSW in 2008. This amounted to approximately 2.04 kg (4.5 lb) of waste per person per day, which was up from 1.68 kg (3.7 lb) per person per day in 1980. Most of the 250 million tons of waste, or 54% (135 million tons), was landfilled, and the rest was recycled or composted. Many states are running out of landfill capacity and are not siting new landfill facilities. These states are taking their solid waste to other states for landfilling. According to the Texas Commission on Environmental Quality (TCEQ), Texans generated 44.8 million tons (98.8 billion lb) of MSW in the year 2000. Approximately 64% of it was disposed of in landfills and 35% was recycled and composted. Based on the available data, we can predict that in Texas, as well as in many other states in the United States, landfills will remain a necessary method of MSW disposal for the foreseeable future. In Europe, more than 70% countries are depositing more than 50% of their waste into landfills.

The major problems associated with landfills as a final waste destination are as follows:

1. **Loss of materials:** Landfills only offer an "open-loop" waste management cycle (Figure 2.31), which means that significant resources that could be used are instead buried. The lost opportunity of reusing resources using the open-loop system increases the need for virgin material extraction, which is suicide for a circular economy. One reason that landfill mining is increasing in importance globally is that it escalates the reuse of materials (Figure 2.32).

2. **Loss of renewable energy potential:** Organic carbon deposited in landfills is converted by microbes to carbon dioxide and methane. If the methane is not captured, it contributes to climate change (21 times more effectively than CO_2 on a per-mass basis over a 100-year time horizon). However, if captured and used to power turbines for electricity generation or natural-gas-powered appliances or vehicles, the methane becomes a renewable energy resource that can be used to reduce the use of fossil fuels.

The opportunities presented by sound lifecycle approaches to reusing materials are currently being demonstrated. Mining existing landfills creates opportunities to reduce the burden of mining for virgin materials and enhances urban sustainability. The pressure on virgin materials (as presented in Figure 2.32) and mining for virgin materials could be significantly reduced if materials buried in the landfills can be reused or materials are diverted from landfilling at the outset.

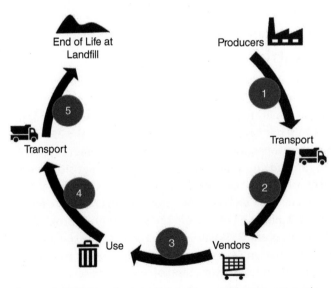

Figure 2.31 Open-loop solid waste management cycle.
Source: Based on Hossain et al. 2014.

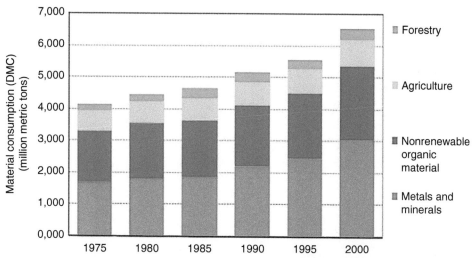

Figure 2.32 Materials consumption in the United States by sector of origin, 1975–2000. *Source*: Modified from Rogich et al. (2008).

3. **Increase in Population and Lack of Available Space for New Landfill Sites:**
   Municipal solid waste landfills often require large tracts of land in or immediately beyond the urban growth areas. Typical landfills may occupy an area
   ranging from several acres to hundreds of acres. Due to the rapid growth and
   urbanization of cities beyond their original city limits, optimizing landfill spaces
   within or close to the city limits is becoming a huge issue for MSW management.
   The following photos (Figure 2.33) demonstrate the idea of urbanization
   and the need for additional space for housing, office space, and commercial
   development in urban areas. Finding a new location for building a new landfill
   every 20 years is next to impossible in certain urban megacities.

4. **Waste minimization and increasing the capacity of existing landfills are**
   becoming major considerations for state agencies and federal regulatory bodies:

   Minimizing the generation of waste and utilizing existing landfill spaces need to
   be considered important parts of waste management to avoid serious crises, as there
   is limited space in urban areas for new landfills. For example, the Naameh landfill,
   which served half of Beirut, Lebanon's population, was closed, and the main waste
   management company, Sukleen, stopped collecting rubbish because it had no place
   to dispose of it. The garbage piled up on every street corner in Beirut, and its stench
   became unbearable. Many of the residents burned their trash on the street corners,
   which was both unhealthy and a serious safety concern. They were in this situation
   primarily because (i) the old dumpsites or landfills were reaching their maximum
   capacity, and the government or city officials were unable to find a place for new ones;

Figure 2.33 Development of major urban cities in 20 years (before and after scenarios). (a1/a2) Dubai, UAE. *Source*: Prasanaik / Wikipedia Commons / Public Domain. (b1/b2) Seoul, South Korea. *Sources*: Don O'Brien / Flickr; travel oriented / Flickr.

(ii) the city officials failed to begin searching for new landfill sites a minimum of three to four years before the existing site reached its maximum capacity so that when the old dumpsites were closed, the new landfills could start accepting garbage; and (iii) finding a suitable location for a landfill close to a major metropolitan city is becoming more challenging every year. This is true not only for Lebanon but also for many other developing and developed countries. *Space* is a big issue.

5. **Social acceptability of new landfills in any community (both developed and developing countries):**

Demonstrations and protests raged against the opening of a new dump (landfill) on the slopes of Mount Vesuvius on 21 October 2010 and led to a riot and violent clash between the local residents and police (Figure 2.34).

The major challenges of how to manage materials demand a new level of awareness and cooperation within and among nations. Domestically, it is important that both the

Figure 2.34 Piles of waste on the road in Naples (https://www.understandingitaly.com/camorra.html).
Source: LHOON/Flickr.

federal and state governments make systematic efforts to enable and encourage all of the sectors of society to ensure that materials are used more effectively and efficiently with less overall environmental toll, thereby enhancing urban sustainability. Ultimately, however, there needs to be an integrated and sustainable waste management system that controls solid waste in a safe, sustainable way and creates healthy urban cities across the globe.

2.3.2 Problems with Open Dumping – Only in Developing Countries

The practice of open dumping of waste causes serious strains on the environment, public health, and the economy by causing water pollution and air pollution that negatively impact the health and safety of the waste workers and the people living near the dumpsites. Some of the major problems associated with open dumping are presented in the following.

2.3.2.1 Water Pollution

Managing generated leachate is a major concern, as dumpsites do not usually have a leachate collection or management system. Leachate seeps through the side slopes and through the soil into the nearby bodies of water, polluting the sources of both surface and underground water. Without any bottom liner, the contamination of nearby water bodies continues throughout dumpsite operation and even after dumpsite closure (Figure 2.35). Surface runoff of rainwater without proper cover system will also contaminate the nearby water bodies. Therefore, the dumpsite remains a major source of contamination for the soil, groundwater, and nearby water bodies.

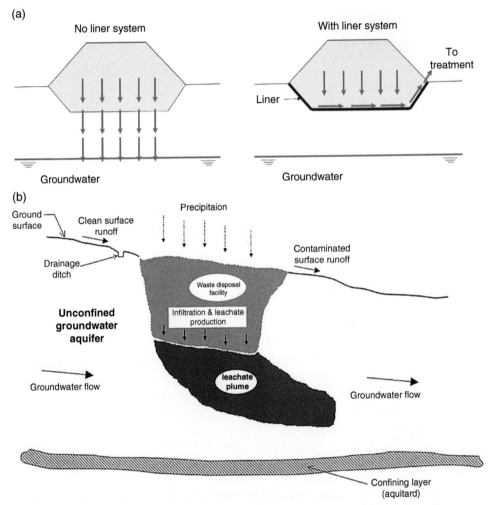

Figure 2.35 (a) Leachate flow with and without liner system. (b) Leachate flow from dumpsite to groundwater (UNEP 2015).

2.3.2.2 Air Pollution

Most open dumpsites do not have a gas collection system and are a major source of greenhouse gas (methane and carbon dioxide) emissions. It has been reported that methane emissions from dumpsites (Figure 2.36) account for 11% of the total methane emissions worldwide (Global Methane Initiative 2018). This makes dumpsites the third largest source of global anthropogenic methane. It has been forecasted that by 2025, if the current scenario continues, 8–10% of the global anthropogenic greenhouse gas (GHG) emissions will be caused by dumpsites (ISWA 2016). The use of cover soils can minimize escape of methane gases and minimize fire potential. However, most open dumpsites do not use daily soil cover as their operational practices.

Some of the evidence of the adverse effects of uncollected waste or unsanitary dumpsites on the health of the environment and nearby residents are highlighted in the following.

- A recent study of toxic waste sites in three countries (India, Indonesia, and the Philippines) concluded that living near a toxic waste site results in exposure that could cause diseases, disabilities, or early death that could be equated to a loss of approximately 829 000 years of good health. In comparison, malaria in these countries, whose combined population is nearly 1.6 billion, causes the loss of 725 000 healthy years (Chatham-Stephens et al. 2013).

Figure 2.36 Uncontrolled burning of waste at open dumpsites (https://cwm.unitar.org/cwmplatformscms/site/assets/files/1457/mia_workshop_istanbul_bangladesh.pdf).
Source: U.S. EPA.

- In another study that was conducted to quantify the health effects on those residing near a waste disposal site in developing countries, airborne dust samples were collected from the Jaleeb Al-Shuyoukh landfill, one of the largest landfills in Kuwait, and a control residential area in Jahra city (Schrapp and Al-Mutairi 2010). The hygienic survey indicated a high amount of airborne dust, bacteria, and fungi within the breathing zone of the residences near the landfill (10–10 000 times higher than in the control area). The high bacterial concentration in the dust samples in the area surrounding the landfill could explain the significant number of dermatological problems such as acne, warts, cysts, and itching skin.

- In a study performed by Kimani-Murage and Ngindu (2007) of the Dandora municipal waste dumping site in Nairobi, Kenya, environmental samples of the soil and water were analyzed, and children living near the dumps were medically examined to determine the content and concentrations of various pollutants (heavy metals, polychlorinated biphenyls, and pesticides) in their bodies that are known to affect human health. Soil samples from the dumpsite were compared to samples taken from another site, Waithaka, which is a peri-urban residential area on the outskirts of Nairobi. Blood investigations confirmed that 50% of the children had low hemoglobin levels and 30% had size and staining abnormalities (microcytosis) of their red blood cells (iron deficiency anemia), a condition brought about by heavy metal intoxication. Further, the blood film studies indicated that 52.5% of the children had marked eosinophilia (an increase in the number of white blood cells mostly associated with allergic reactions), a condition that could lead to chronic rhinitis (irritation of the nasal cavity), asthma, allergic conjunctivitis, and dermatitis. Additionally, the soil samples analyzed from locations within and adjacent to the dumpsite showed high levels of heavy metals emanating from the site – in particular, lead, mercury, cadmium, copper, and chromium.

- e-Waste has become a major source of concern in developing countries recently. A study conducted by Greenpeace in the electronic recycling yards in Delhi in 2005 clearly indicated the presence of high levels of hazardous chemicals, including deadly dioxins and furans in the areas where this primitive recycling takes place (Network 2002). Scientists who have examined Guiyu, China, (one of the popular destinations of e-waste recycling activities) have determined that because of the waste, the location has the highest levels of cancer-causing dioxins in the world. Pregnant women are six times more likely to suffer a miscarriage, and seven out of ten kids have too much lead in their blood. There is a paucity of data on the burdens of heavy metal exposure on the human body in developing countries.

2.3.2.3 Safety and Operation

The safety of waste pickers is greatly compromised in the working face due to poor working conditions and the lack of safety measures. When a trash truck enters a dumpsite, hundreds of waste pickers run toward or alongside it. This seems unbelievable, but many view it as comparable to the hordes of people who chase a limo that contains a celebrity, hoping to get a glimpse of them. Ironically, the waste pickers' "celebrity" is the trash that has been thrown away by people for whom it had no value. They run because they want to retrieve as many recyclables or materials of value as possible, so that they can sell them in the market. It is their only means of making a living, and they sometimes jump onto the truck, or even let the trash be dumped on them to get materials of value before anyone else. These are clearly unsafe practices that result in waste pickers dying every day around the globe.

A study conducted in Bahir Dar City in Ethiopia showed that almost 64% of their waste workers are injured more than once every year (Gizaw et al. 2014). The poor management of dumpsites is a major concern in developing countries. They are very poorly managed, and most of the wastes are dumped without proper compaction and without any specific layout or plan. In many cases, there is no trash compactor at the site. The trucks dump the waste onto the dumpsite and a loader just spreads the waste onto the working face without having performed any compaction. The poor or lack of compaction allows easy passage of oxygen into the waste that causes fires to break out in the landfill. Additionally, the poor management of the side slopes of the landfill and the lack of maintaining appropriate slopes or compaction can be responsible for landfill or dumpsite slope failures. Many fatal or catastrophic slope failures have occurred in landfills in recent years, including the Koshe dumpsite failure in Addis Ababa, Ethiopia, which caused the loss of more than 150 lives.

2.3.2.3.1 CASE STUDY: Koshe Open Dumpsite Failure in Addis Ababa, Ethiopia – March 2017

On 11 March 2017, at 8:00 p.m. local time, a waste slope failure occurred at the Koshe waste dumpsite when a mass of waste collapsed onto a slum that was built close to the toes of the dumpsite slope and killed at least 120 people in the city of Addis Ababa, Ethiopia (livescience.com, 17 March 2017). Based on the information provided by the corresponding city officials and a site visit, the waste height had been approximately 40–50 m (120–150 ft.), with steep side slopes (slope angles between 60° and 70°). The solid waste management and landfill experts from the University of Texas at Arlington (UTA) and Solid Waste Institute for Sustainability (SWIS), United States, traveled to the failed site to investigate the waste slide area and collect pertinent information. A summary of the Koshe dumpsite slope failure, a stability analyses, and recommendations for avoiding future failures and minimizing damage and/or saving people who live next to dumpsites is presented in the following.

2.3.2.3.2 Background of Reppie (Koshe) Open Dump Site

The Reppie Solid Waste Disposal Site (SWDS), also known as "Koshe," is an open-air dumpsite located about 13 km (8 miles) southwest of the Addis Ababa city center, in the Kolfe-Keranyo sub-City area (Figure 2.37). The landfill's geographical coordinates are latitude 8° 58′ 29.6754″ N (or 8.9749° N) and longitude 38° 42′ 43.6314″ E (or 38.7121° E). The solid waste is dumped on the fringe of the Tinshu Akaki River, and the dumping area covers 364 000 m² (or approximately 36.4 ha). The site is on the northeast side of the new Addis Ababa Ring Road, heading toward the western parts of the city.

The Addis Ababa city administration is responsible for the full operation of the currently active SWDS. Koshe has been acting as an unplanned open dump since its inception in 1968, and waste is dumped at the site without prior processing or treatment. According to the Solid Waste Recycling Disposal Project Office (SWRDPO), there is approximately 3.5 million tons (7.7 billion lb) of waste at the site, with daily additions of 3000 tons (6.61 million lb). Households account for about 76% of the waste, with the other 24% generated from institutions, hotels, industries, and street sweeping.

The total area of the Koshe dumpsite is 364 000 m² (36.4 ha). Eighteen hectares have been closed to capture and flare landfill gas (LFG), 70 000 m² (7 ha) is used for the waste-to-energy facility, and the remaining 114 000 m² (11.4 ha) is open for receiving the city's waste.

The dumpsite closed its operation in December 2015, and the city's waste was then dumped in a newly designed sanitary landfill in Sendafa, which is just outside the city of

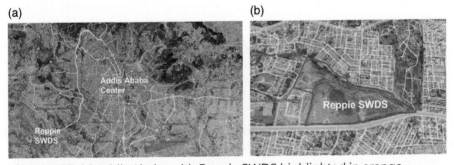

(a) (b)

Figure 2.37 (a) Addis Ababa with Reppie SWDS highlighted in orange. *Source*: Google Earth. (b) Reppie SWDS in context with surrounding neighborhoods and Ring Road. *Source*: Google Earth.

Addis Ababa. However, due to opposition from the Sendafa residents, the city officials again opened the closed portion of the Koshe landfill for dumping waste in April 2016. It should be mentioned that the Koshe dumpsite was already beyond its capacity; hence, additional activities at the site could cause pressure on the dumpsite, which has steep side slopes.

2.3.2.3.3 Description of a Cross Section and Causes of Failure

The steep slope section (slope angle between 60° and 70 °) at the corner of the dumpsite has not yet failed and is presented in Figure 2.38. Figure 2.39 shows a close-up of the slope failure area and scarp at the Koshe dumpsite. The waste mass slid over the residents living on the side slope and those living close to the toe of slope of the dumpsite. The failed slope was as steep as the slope depicted before failure in Figure 2.38.

Before the failure, the side slope was about 60°–70°, which is considered very steep, compared to the allowable designed waste side slopes of 20°–25° in the United States. Based on the information provided, it appears that the city has been dumping waste of different compositions at the site for the past 50 years. Varying degrees of decomposition of solid waste at different depths were clearly visible during the site visit. The schematic cross section of a typical waste side slope is presented in Figure 2.40.

Figure 2.38 Steep slope of the Koshe dumpsite.
Source: Photo taken by author during SWIS site visit.

(a) (b)

Figure 2.39 (a) Slope failure at Koshe dumpsite. *Source*: Photo taken by author during SWIS site visit. (b) Waste mass that slid over nearby residents. *Source*: Photo taken by author during SWIS site visit.

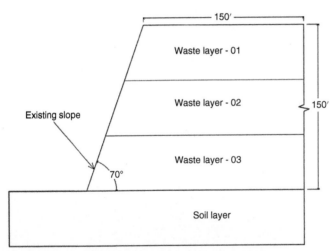

Figure 2.40 Schematic cross section of the failed slope.
Note: 1 ft. = 0.3048 m.

In addition to the slope failure, cracks were also visible along and parallel to the failed slope surface (Figure 2.41). Cracks are an indication of the ongoing movement of the slope and a warning that there is potential for near-future failure along and beyond the existing failed slope area. This is a major cause of concern because during the site visit, the SWIS team found that local residents are still living at the corners of the dumpsite, and if the corner slope fails, there is potential for more lives to be lost.

(a) (b)

(c)

Figure 2.41 (a–c) Visible cracks at top of dumpsite (parallel to failed slope).
Source: Photo taken by author during SWIS site visit.

Based on the preliminary site investigation and assumptions of waste parameters, slope stability analyses were performed for the waste slopes at the Koshe dumpsite, using the computer program, STABL. The results are presented in Figure 2.42. The results of the slope stability analyses clearly showed that the factor of safety for the slope was less than 1 in all of the cases that were analyzed. (The minimum design factor of safety for the slope should be greater than 1.5.)

They also showed that with the inclusion of additional load at the top of the site (because of the waste dumping at the site), the factor of safety was reduced to 0.45 in the steep slope Figure 2.43a. It is very interesting to note that with the same waste

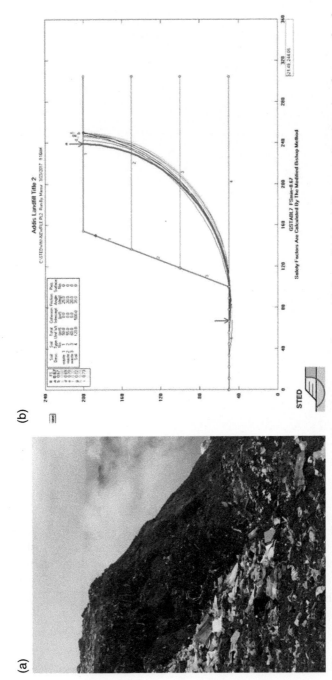

Figure 2.42 Slope stability analyses. (a) Slope that experienced failure. (b) Modeling results showing factor of safety less than 1.

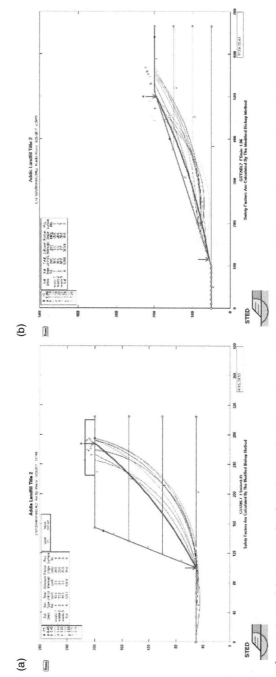

Figure 2.43 (a) Factor of safety is 0.45 in steep slope with load, because of the waste dumping activities. (b) Factor of safety is more than 1 for 3:1 slope.

composition, the factor of safety increased to more than 1.0 with 3H:1V slope that is the minimum side slope typically used in developed countries (Figure 2.43b).

The causes of the slope failure of the Koshe dumpsite presented in the following are based on the preliminary site investigations, visual inspections, and preliminary slope stability analyses results.

1. **Steep Side Slopes**

Landfill side slopes are expected to be 3H:1V (3 horizontal to 1 vertical), which translates to side slopes of less than 20°. Before failure, the side slopes at the failed locations and at the corner of the currently active location were 0.4H:1V (60°) to 0.33 H:1V (71°). The slope was too steep for any kind of solid waste to be stable, which inevitably led to failure. Combined with the second causes of failure, which are described later, the magnitude of the slope failure became catastrophic.

2. **Degradation of Waste Materials and Loss of Strength with Time**

The waste at the Koshe dumpsite is composed of more than 60% food waste and is primarily organic in nature. Organic waste decomposes quickly, leaving plastics and other nondegradable components as the primary waste composition. As the percentage of plastics increases with the decrease of other waste components, the strength of the solid waste is expected to decrease significantly. This phenomenon often causes slope failure, and it was clearly one of the main reasons for the slope failure at the Koshe dumpsite.

3. **Additional Loads on the Top and the Side Slopes Because of Renewed Waste Dumping After Closing the Site**

During the site visit, it was clearly visible that an extra load on top of the dumpsite (Figure 2.38) caused additional stress on the side slopes. Additional cracks (presented in Figure 2.41) were also observed across the site because of the newly resumed dumping activities. This is an indication of additional pressure on the dumpsite and continued risk of further slope failure. If it rains continuously for a few days at the dumpsite, the cracks will act as an easy pathway for water intrusion and will lead to an increased driving force and decreased shear strength of the waste mass, ultimately leading to another failure of the waste side slopes.

2.3.2.3.4 Remedial Measures

The immediate action plan presented here for avoiding further catastrophic failure of the waste mass is based on the site visit (Figure 2.44), the results of the stability analyses, and conversations with the city waste management officials. The preliminary action plan can be modified based on further site investigations, waste characteristics, and availability of a buffer zone.

Figure 2.44 Additional waste piled on top of waste slope.
Source: Photo taken by author during SWIS site visit.

1. Relocate residents living next to the toe of the slope at the corner of the dumpsite (Figure 2.39). If the residents are not moved and that section of the slope fails, there is potential for further loss of human life. That corner has a high potential for failure, as

   - the slope is very steep at the corner.
   - additional loads are placed on the top of the slope during the dumping activities.
   - cracks are visible around the corner.

2. Create a 50 m to 100 m (164 ft. to 328 ft) buffer zone from the toe of the waste slope at the Koshe dump site to prevent further loss of human life.

3. Stop the waste dumping operation at the Koshe dumpsite as soon as possible and dump the waste at some other place. The newly designed sanitary landfill at Sendafa, which is outside the city of Addis Ababa, is a viable alternative.

4. Continue the site investigation at the Koshe dumpsite for closure, reclamation, or possible future uses of the site.

Similar dumpsite failures caused the loss of lives in Sri Lanka, China, and Indonesia in recent years. It is imperative that we learn how to close existing dumpsites without causing sociopolitical problems or creating a situation like that which occurred in Lebanon in 2015.

2.4 COMPOSTING

Many developed and developing countries are actively working to divert organic waste from the landfills/dumpsites and use them for composting. In developed countries, source separation of organic waste from mixed waste makes it relatively easy to divert solid waste from final disposal through composting. However, in developing countries, the main challenge still remains, as source separation of organic waste has been marginally successful. Therefore, to make composting part of a sustainable and viable solution for waste management, source separation of organic waste needs to be seriously considered.

2.5 RECYCLING

Recycling is one of the major components of a sustainable waste management system. Recycling depends heavily on waste collection operations (source-separated or mixed waste collection).

Recycling is relatively easy in developed countries, as the materials are often source separated, and collecting and transporting them to a material processing center is easy. However, the recycling rates vary significantly from one country to another and one continent to another, even in the developing countries. The recycling rate for MSW in developing countries is presented in Figure 2.45 (World Bank 2018a).

The recycling rate has been between 20 and 30% in the United States for the past 20 years. The market for selling recycled products is volatile, and unless the selling prices are stabilized, there is always going to be a major issue. Currently, without federal or state governments' direct intervention and/or subsidies, the recycling market is not in a position to survive, as the rate of recycling is going to decrease significantly. With the recent (2018) Chinese restrictions on recycling materials, the market is on the verge of collapsing, and there are no opportunities to make money from the products that are piling up in storage places.

Data indicates that many cities in the United States are finding it difficult to maintain their recycling programs, as they too have collected or sorted recycled materials sitting in storage places without a potential market to sell them. This is happening across the country because of the previous heavy reliance on the Chinese market for recycling materials and the lack of a plan for creating local markets for the products. In 2016, the United States exported $11 billion of recycling materials to China, of which

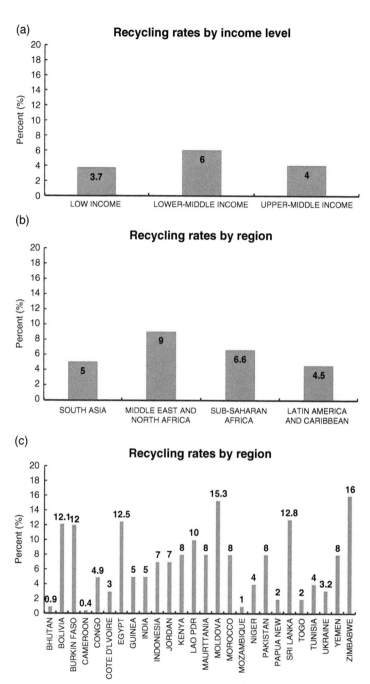

Figure 2.45 MSW recycling rates in developing countries. (a) Recycling rates by income level. (b) Recycling rates by region. (c) Recycling rates by countries (Redrawn from World Bank 2018a).

approximately $1.2 billion were plastics. The United States expected to export plastics worth $300 million to China in the first quarter of 2018, but because of the restrictions and strict recycling regulations, only $9 million of plastic materials were exported. The impact of the Chinese restrictions, which have now expanded to a ban, on recycling products is so significant globally that the recycling market is at risk of being eliminated or reduced to the level that most recycled materials may end up going to dumpsites or landfills. Many cities and waste management business operators, even in developed countries, are being forced to come up with ways to use recycled materials locally or find new markets for them. Otherwise, they will end up in dumpsites or landfills.

In developing countries, recycling is mostly done by the informal sector where most of the recyclable materials are taken by waste pickers (Figure 2.46) before they even reach the waste management facilities or open dumpsites, and those that do reach the dumpsite are quickly confiscated by the waste pickers. So, sometimes almost 100% of the recyclable materials are taken out from the working phase.

Sweden that has an excellent recycling program ran out of garbage for their waste to energy and were forced to import waste from their neighbors, mainly Norway and England (Hickman 2018). The incineration of solid waste provides power to 250 000 of their homes and heat to 810 000. Swedes are so big on recycling that less than 1% of all household waste generated in the country has been landfilled annually since 2011, which is why they ran out of garbage that they needed for the waste to energy plants. The World (2012) reported that Sweden imports about 800 000 tons (1.76 billion lb) of trash annually from other regions in Europe to run their power plants.

Figure 2.46 Waste pickers picking recyclables from a dumpsite.
Source: Francisco Magallon/Wikipedia Commons/Public Domain.

2.6 WASTE-TO-ENERGY (WTE)

WTE plants are common in many developed countries, especially in Europe, as they generate heat or electricity that is used for heating their homes during the winter. This is a very good technology for European countries, as the heating value for their waste stream is at the higher end. Based on the data of the waste characteristics presented in Figure 2.47, the content of the waste in European countries consists of a lot of paper and plastics and very few organic materials.

The moisture content of the waste in many European countries is very low compared to that of developing countries. The success of the WTE plants in developed countries, and especially Europe, has motivated many developing countries to try to use their waste-to-energy plants to manage their MSW. Unfortunately, they have had marginal success, as the organic and food waste components are very high in their waste stream, which translates into a high moisture content. In some cases, the WTE facilities have operated for a very short time (from a few days to a few months) before completely shutting down. The following section is presented from a news article published by the author in The Daily Observer, "Waste to Energy (WTE): Will it Work in BD?" – Bangladesh, September 2020 (Hossain 2020).

MSW or garbage generation is increasing rapidly because of the increasing population in developing countries. Sustainable waste management is a major challenge for both megacities like Dhaka with 20 million people and smaller cities/rural areas. The availability of land/space for building waste management systems (landfills, composting, recycling, and/or waste-to-energy) can constrain the

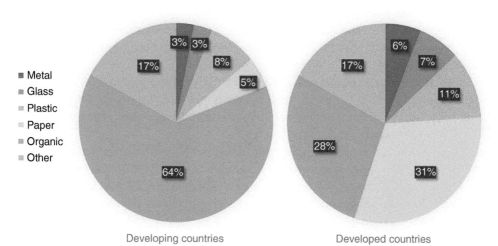

Figure 2.47 Waste composition in developing and developed countries (Redrawn from World Bank 2018a).

decision making. However, we need to pay special attention to (i) availability of land/space and (ii) applicability of technology in a specific region (based on waste characteristics).

Waste-to-energy or incineration can provide a lucrative solution to waste management, as WTE plants may address the issue of land and space in both developed and developing countries. WTE, or incineration systems, require less land or space and can process a large volume of waste in a single processing plant. It is an appropriate technology for waste management in developed countries, such as those in Europe, the United States, Japan, and South Korea, as their waste is relatively dry and the presence of a large number of plastics, paper, and other combustible materials makes it a good source of energy.

The Bangladesh Constitution was amended in 2011 to include a directive to the state to protect the environment and natural resources for current and future generations. The current decision of adopting waste management through WTE or incineration directly opposes the policy of Prime Minister Sheikh Hasina's vision and plan for a sustainable urban environment for Bangladesh. This write-up highlights major issues that need in-depth feasibility studies and evaluations when considering WTE in developing economies and Bangladesh:

1. Applicability: Waste in developing countries like Bangladesh contains more food (in excess of 70%) and moisture than that of European countries, which makes it neither applicable nor cost effective to generate power through WTE or incineration due to its low calorific value. However, following the success of the WTE plants in Europe and intense lobbying from many WTE companies, many developing countries, including Bangladesh, are trying to use the WTE technology without understanding the applicability of the technology or the serious environmental and public health consequences. WTE is not an appropriate technology for Bangladesh.

2. Cost: The initial capital investment and the operating costs make it very expensive to operate a WTE plant. Moreover, the technology does not apply to the type of waste that Bangladesh and other developing countries are dealing with.

3. Environment and public health concerns: Burning the waste at WTE plants generates 20% bottom ash. In most cases, there is no plan for managing the bottom ash, and it ends up in water bodies, polluting the nearby rivers and creating a serious problem for clean water and agricultural lands.

It is difficult to enforce air pollution regulations and control the air pollution that results from the toxic emissions from WTE plants and incinerators. They are a source of particulate matter (PM 2.5), which are tiny particles of dust that can lead to decreased lung function, irregular heartbeat, heart attacks, and premature death. On 23 March 2019, South Delhi residents organized one of the largest open-chain rallies to protest the Okhla waste-to-energy plant. Their complaint against the plant that generates electricity in the midst of more than one million residents who are living in colonies is that it is filling the atmosphere with stench and spewing toxic fumes that are making people ill. (Sambyal et al. 2019). "The number of asthma patients admitted to emergency rooms and intensive care units has gone up since the plant was set up," said Shailendra Bhadoriya, a consultant cardiologist with the Fortis Escorts Heart Institute, one of the three hospitals in the plant's vicinity. The Okhla plant, like most WTE plants in the country, has remained controversial since its inception. Despite a public interest petition being filed in the Delhi High Court against the plant proposal in 2009, it was commissioned in 2011.

4. Success in developing countries: In 1987, the old Timarpur WTE plant in Delhi, India, had to shut down just 21 days after opening due to operational failure brought on by a low calorific value of incoming waste (2.5–2.9 MJ/kg) (Shah 2011). Since 1990, 14 WTE plants have been installed in India, but half of them have been closed, and the remaining ones are under scrutiny (Sambyal et al. 2019). Along with the low calorific value and high moisture content of the MSW, the presence of inert materials has also been the cause of several projects failing in developing nations.

Only 20% of thermal WTE plants worldwide are located in developing countries (UNEP 2019). For instance, in India, the construction and demolition waste in mixed MSW resulted in operational failure of several thermal WTE plants (Planning Commission, India 2014). Other reasons for failure are the lack of interest from investors and the public sector, public outcries against plant locations, and poor quality and quantity of waste. The Reppie thermal WTE plant in Addis Ababa had to terminate its operations just weeks after its inauguration due to lack of funding (Mulugeta 2019). The deficit of installed capacity, lack of suitable pollution control and monitoring standards, and the absence of controlled landfills for ash disposal have been recognized as key factors for the limited success of WTE plants in developing countries (UNEP 2019).

The choice of technology in managing solid waste is a major issue in developing countries, and the wrong technology is often chosen because of the lack of trained or highly skilled waste professionals. It is important to the success and sustainability of the waste management systems in developing countries to have guidelines that will assist them in selecting and applying the appropriate technology for managing their systems.

2.6.1 Case Study: Reppie Waste-to-Energy (WTE) Plant in Addis Ababa, Ethiopia

Ethiopia belongs to the category of low- and middle-income countries (LMIC).[1] Since 1990, the population has doubled, to reach its current number of 110 million. The GDP has increased sixfold, and the per capita GDP has tripled. It has met the major targets of the Millennium Development Goals (MDGs), which are a 100% increase in

---

1 https://wellcome.ac.uk/grant-funding/guidance/low-and-middle-income-countries.

school enrollment, a 50% reduction in poverty rates, and an increase in life expectancy from 47 years to 66 years. But despite 30 years of remarkable achievements, it continues to face numerous challenges. A primary driver of the trade deficit is energy, and since the turn of the century, Ethiopia's trade deficit has grown to more than $12 billion. However, fossil fuel imports, which currently account for 7% percent of the nation's primary energy usage, consume 80% of the export earnings (National Bank of Ethiopia 2016/2017).

Imported oil and electricity from renewable sources account for 8% of Ethiopia's energy use. The remaining 92% comes from traditional sources, which are mainly charcoal, wood, and dung. According to FAO (FAO 2017), Ethiopia is among the top three charcoal producing countries in the world: the second largest in Africa. Ethiopians use more than four million tons of charcoal (120 million GJ[2] of energy) each year. Charcoal is produced in a traditional kiln and requires a minimum of four tons of wood to produce a single ton of charcoal (Girard 2002). Thus, to produce charcoal, 16 million tons of wood (20 million cubic meters) are burned every year, which amounts to one million hectares of deforestation annually (FAO 2001). Deforestation increases carbon dioxide emissions, a major GHG that is contributing to climate change and global warming. Moreover, burning four million tons of charcoal releases an additional four million tons of carbon dioxide to the atmosphere over the period of a year (Girard 2002). The Ethiopian government understood these challenges even years ago and promoted the use of kerosene by offering it at a subsidized rate to discourage charcoal production. When oil prices soared in the first decade of the century, however, the subsidy became financially unsustainable and the government ended the program (Reuters 2008). Now, electricity is sold at subsidized rate of $0.02 per kWh while the production cost is $0.09 per kWh. Consequently, electricity use has tripled.

Since 1990, Ethiopia has added, on average, two million citizens and $2.5 billion of GDP/yr. At the same time, energy use has increased at a rate of about 40 peta Joule PJ/yr. At these rates, by 2030, the population will be 133 million and the GDP and GDP per capita will be close to $114 billion and $900 billion, respectively. As the result, energy use might increase by almost 500 PJ (a million GJ).

If Ethiopia continues to rely on traditional fuel and imported oil to meet its growing energy demands, fossil fuel imports will continue to devour a significant portion of their export earnings, and charcoal production will continue to increase both deforestation and GHG emissions. Hence, the negative impact on the economy and the environment will persist. Therefore, there must be a paradigm shift toward sustainable energy use to cover Ethiopia's particular needs.

Currently, electricity that is primarily derived from hydropower is the only sustainable energy source. As a result, electricity consumption increased by almost

---

2 Giga Joules (GJ) is a billion joules.

300% between 1990 and 2010. Despite such remarkable growth, however, Ethiopia's electricity consumption per capita (250 MW) is much lower than the average for sub-Saharan African countries (1750 MJ).[3] According to the World Bank,[4] the urban population of Ethiopia in the first decade of the century was about 15 million, which was one of the lowest in the world and significantly below the average for sub-Saharan Africa (37%). It is expected to grow at a rate of 5.4% per year to reach 30% by 2030. Addis Ababa, the capital city, is home to 25% of the urban population and is the fastest growing city in Africa in terms of population and economy.[5] The population grew exponentially from 392 000 in 1950 to over 3 million at a rate of 3.6%. At this rate, it is likely that the population of 4.5 million in 2020 may surpass six million by 2030.

Nearly 75% of the household waste generated in Addis Ababa is organic. About 17% is recyclable materials (mainly paper and plastic), and the remaining 8% is character-ized as hazardous waste. Households generate more than 75% of the waste, followed by public and commercial institutions (18%) and street cleaning (6%). The city reported that about 0.5 kg of waste is generated per person per day, and it manages to collect 2000 tons per day, which is less than 90% of the waste. They estimate that about 65% of the municipality solid waste is organic; about 15% of the waste is recyclable, such as paper and plastic; and the remaining 20% is composed of electronic waste and haz-ardous waste from hospitals and industries.

The MSW has been disposed of for more than fifty years on an open dump site within Addis Ababa that is famously known as Koshe or Reppie. The city tried to close the dump site properly by capturing methane and implementing a flaring program that qualified for Clean Development Mechanism (CDM) while they built a sanitary landfill outside the city in the Oromia region in a place called Sendafa. The Reppie waste-to-energy plant was set up at the Koshe open dump site to incinerate the waste and produce 50 MW of electricity. The Sendafa landfill was built with financial support from the French Development Agency (AFD) and French developers and met all of the technical requirements for a modern sanitary landfill. The project failed to under-take a proper social and environmental impact assessment, however, and as a result became a target during the 2016 civil unrest, and the city was forced to abandon the project.

The Reppie WTE plant was commissioned in 2014 and became operational in 2019. Typically, WTE plants generate 500–600 kWh of electrical energy per ton of

---

3 https://data.worldbank.org/indicator/EG.USE.ELEC.KH.PC?locations =ZF.

4 https://openknowledge.worldbank.org/bitstream/handle/10986/22979/Ethiopia000Urb0ddle0 income0Ethiopia.pdf?sequence =1&isAllowed = y.

5 https://documents.worldbank.org/curated/en/559781468196153638/pdf/100980-REVISED-WP-PUBLIC-Box394816B-Addis-Ababa-CityStrength-ESpread-S.pdf.

waste incinerated, but the Reppie plant produces on the average 352 kWh of energy per ton of waste. Consequently, its power capacity has been reduced to 21 MW as opposed to 50 MW due to the nature of the waste available. The plant utilizes 15% of the energy produced from their operation, making the net production of energy around 18 MW if both the engines are running. During the author's recent visit to the Reppie WTE Plant on 01 March 2021 (Figure 2.48), only one engine was running for energy production, which made the net energy production for that day to 8.5 MW.

There are many reasons for the lower-than-expected production of energy:

1. The city estimates that about 65% of the municipality solid waste is organic, and organic waste has a low calorific value. About 15% of the waste is recyclable, such as paper and plastic.

2. The city pays a subsidy of 4 cents (US) per kg for paper products, 8 cents per kg for organic matter to make compost, and 5 cents per kg for plastic bottles, which totals nearly $2.5 million of annual income for those receiving the subsidies and benefitting from the program. Twelve small factories produce plastic pellets that are exported to China, which means that the highly combustible waste such as paper and plastic that the plant depends on is no longer available.

3. The waste they are using comes from two sources: 25% from fresh collected solid waste from cities and 75% from recycled waste from the Koshe dumpsite, which has higher volume of plastics. This seems to be the reason they are able to generate more than 10 MW energy; if they depended only on fresh waste, the energy generation would be even lower.

Since the plant is generating less than 40% of the expected energy and the operating costs are very high, it requires subsidies to stay alive. The city is planning to subsidize the plant to keep it running since it manages to burn nearly half of the municipal waste. In the long run, the sustainability of the plant is questionable. The WTE plant is also dumping their daily produced highly toxic bottom ash at the Koshe dumpsite, which is a very unhealthy practice as the waste workers are retrieving metal from the bottom ash without being aware of the negative health implications. The air quality data from the Reppie WTE plants was not available during the author's visit to the facility.

Addis Ababa, despite the fact it has invested heavily on advanced waste management technologies, faces a serious waste crisis. The Koshe dump site is operating beyond its capacity and is accepting more than a 1000 tons of waste daily. This cannot continue in the long term. There is clearly a need for a robust waste management program to address the eminent waste crisis in the short term, and the sustainability challenge in the long run is clear.

(a)

(b)

(c)

Figure 2.48 (a–c) SWIS visit to Reppie WTE Plant, Ethiopia, in March 2021.

2.7 SUMMARY OF CURRENT CHALLENGES OF WASTE MANAGEMENT IN DEVELOPING COUNTRIES

Solid waste and landfill management are issues not only in the United States but also in European and developing countries around the world. Regions of Africa, Latin America, South Asia, East Asia, the Pacific Islands, and the Caribbean are urgently in need of help in mitigating the problems associated with increasing population growth, urban consumption, and waste production. Without incorporating safe waste management practices into the development plan, the proliferation of poor health and sanitation conditions will persist. The real challenges of successful solid waste management are essentially those listed as follows:

- Lack of public awareness of waste management and its impact on health and well-being, the environment, and the local and global economies.

- Insufficient collection and management systems in developing countries that cause major health, sanitation, and environmental issues.

- Improperly managed open dumpsites and their serious consequences on environment and public health.

- Lack of space for new landfills every 20/25 years as urbanization and migration of population to urban areas causes serious strain on urban space.

- Inadequate sustainable waste management technology and implementation of the technology in developing countries.

- Lack of training on waste management technology and lack of management assistance for local authorities and waste management personnel in developing countries.

- The marginal or nonexistent guidelines or *roadmap* for developing waste management protocols or regulatory frameworks at both the local and national levels in many of the poorest countries in the world.

REFERENCES

Chatham-Stephens, K., Caravanos, J., Ericson, B. et al. (2013). Burden of disease from toxic waste sites in India, Indonesia, and the Philippines in 2010. *Environmental Health Perspectives 121* (7): 791–796.

Eurostat (2009). *Municipal solid waste generation, landfilling, incineration, recycling, and composting from 1995 to 2007. Environment and Energy*. Environmental Data Centre on Waste. http://epp.eurostat.ec.europa.eu/portal/page/portal/waste/data/sectors/municipal (accessed March 2020).

FAO (2001). Forestry Outlook in Africa. http://www.fao.org/3/ab582e/AB582E04.htm.

FAO (2017). The charcoal transition: greening the charcoal value chain to mitigate climate change and improve local livelihoods, (ed. J. van Dam). Rome: Food and Agriculture Organization of the United Nations. http://www.fao.org/3/a-i6935e.pdf.

Ferronato, N. and Torretta, V. (2019). Waste mismanagement in developing countries: a review of global issues. *International Journal of Environmental Research and Public Health* 16 (6): 1060.

Girard, P. (2002). Charcoal production and use in Africa: what future? *Unasylva* 211(53). http://www.fao.org/tempref/docrep/fao/005/y4450e/y4450e05.pdf.

Gizaw, Z., Gebrehiwot, M., Teka, Z., and Molla, M. (2014). Assessment of occupational injury and associated factors among municipal solid waste management workers in Gondar town and Bahir Dar City, Northwest Ethiopia, 2012. *Journal of Medicine and Medical Sciences 5* (9): 181–192.

Global Economy, (n.d.). *World Economy*. TheGlobalEconomy.com. Retrieved 2021, from https://www.theglobaleconomy.com/.

Global Methane Initiative. (2018). *About Methane*. https://www.globalmethane.org/about/methane.aspx (accessed 6 April 2020).

Hickman, M. (2018). Sweden Runs Out of Garbage, Forced to Import from Neighbors. Treehugger. https://www.treehugger.com/sweden-runs-out-of-garbage-forced-to-import-from-norway-4868335.

Hossain, M.D. (2009). Waste management. *The Daily Star* (1 August 2009). https://www.thedailystar.net/news-detail-99617 (accessed 7 February 2022).

Hossain, M.D.S. (2020). Waste to Energy (WTE): Will it work for BD? *The Daily Observer* (29 September). https://www.observerbd.com/news.php?id=277269.

Hossain, S. and Samir, S. (2014). Resource recovery and sustainable material management through landfill mining. Doctoral dissertation. The University of Texas at Arlington.

IOM UN Migration (2015). *World Migration Report*. https://www.iom.int/world-migration-report-2015.

ISWA (2016). *A Roadmap for closing Waste Dumpsites*. International Solid Waste Association. https://www.iswa.org/fileadmin/galleries/About ISWA/ISWA_Roadmap_Report.pdf.

Kimani-Murage, E.W. and Ngindu, A.M. (2007). Quality of water the slum dwellers use: the case of a Kenyan slum. *Journal of Urban Health 84* (6): 829–838.

Mulugeta, T. (2019). Contentious waste power plant to resume operations. *Addisfortune News*. https://addisfortune.news/contentious-waste-power-plant-to-resume-operations.

National Bank of Ethiopia 2016/2017. Annual Report. https://nbebank.com/wp-content/uploads/pdf/annualbulletin/NBE%20Annual%20report%202016-2017/NBE%20Annual%20Report%202016-2017.pdf?x49341.

Network, B. A. (2002). Exporting harm: the high-tech trashing of Asia. http://www.ban.org/E-waste/technotrashfinalcomp.pdf.

Planning Commission, India (2014). *Report of the Task Force on Waste to Energy* (Volume I). http://planningcommission.nic.in/reports/genrep rep_wte1205.pdf.

PRB (2016). *World Population Data Sheet*. PRB. https://www.prb.org/resources/2016-world-population-datasheet/#:%7E:text=PRB%20Projects%20World%20Population%20Rising,Population%20Reference%20Bureau%20.

Reuters (2008). Ethiopia ends fuel subsidy, increase pump prices. https://www.reuters.com/article/us-ethiopia-fuel/ethiopia-ends-fuel-subsidy-increases-pump-prices-idUSTRE49318S20081004.

Ritchie, H. (2018). *Urbanization*. Our World in Data (13 June). https://ourworldindata.org/urbanization

Rogich, Cassara, A., Wernick, I., et al. (2008). Material Flows in the United States. A Physical Accounting of the US Industrial Economy. An WRI Report.

Sambyal, S.S., Agarwal, R., and Shrivastav, R. (2019). Trash-fired power plants wasted in India. *DowntoEarth*. www.downtoearth.org.in/news/waste/trash-fired-power-plants-wasted-in-india-63984.

Schrapp, K. and Al-Mutairi, N. (2010). Associated health effects among residences near Jeleeb Al-Shuyoukh landfill. *American Journal of Environmental Sciences* 6 (2): 184–190.

Shah, D. (2011). Delhi's Obsession with "Waste-to-energy" Incinerators: The Timarpur-Okhla Waste to Energy Venture. http://www.no-burn.org/wp-content/uploads Timarpur.pdf.

UNEP (2015). *Global Waste Management Outlook*. UNEP.

UNEP (2019). *Waste-to-Energy: Considerations for Informed Decision-Making*. UNEP.

UN-Habitat (2016). *World Cities Report 2016: Urbanization and Development – Emerging Futures*. United Nations. https://unhabitat.org/world-cities-report.

United Nations – Department of Economic and Social Affiars (2018). *World Urbanization Prospects*. https://population.un.org/wup/Download.

USEPA (2013). *Non-Hazardous Waste Management Hierarchy*. Environmental Protection Agency. http://www.epa.gov/osw/nonhaz/municipal/hierarchy.htm (accessed 12 December 2013)

Vij, D. (2012). Urbanization and solid waste management in India: present practices and future challenges. *Procedia-Social and Behavioral Sciences 37*: 437–447.

World Bank (2018a). *What a Waste: An Updated Look into the Future of Solid Waste Management*. https://www.worldbank.org/en/news/immersive-story/2018/09/20/what-a-waste-an-updated-look-into-the-future-of-solid-waste-management.

World Bank. (2018b). *Population, total | Data*. https://data.worldbank.org/indicator/SP.POP.TOTL.

World Bank Group (2012). *What a Waste: A Global Review of Solid Waste Management*. https://siteresources.worldbank.org/INTURBANDEVELOPMENT/Resources/336387-1334852610766/What_a_Waste2012_Final.pdf.

World Bank (2021). *Data Catalog. World Development Indicators*. Retrieved from https://datacatalog.worldbank.org/search/dataset.

Chapter 3

Case Studies – SWIS Winter School Ambassadors

The Waste Crisis: Roadmap for Sustainable Waste Management in Developing Countries,
First Edition. Sahadat Hossain, H. James Law and Araya Asfaw.
© 2022 John Wiley & Sons Ltd. Published 2022 by John Wiley & Sons Ltd.

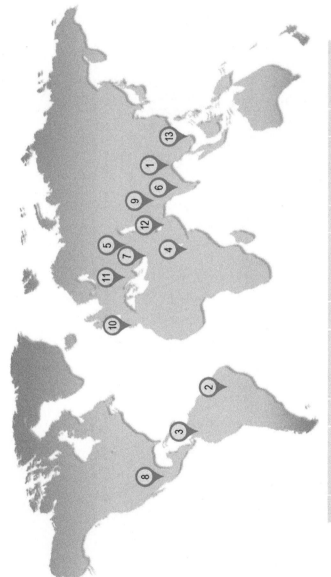

1	Bangladesh	4	Ethiopia	7	Lebanon	10	Portugal	12	UAE
2	Brazil	5	Georgia	8	Mexico	11	Serbia	13	Vietnam
3	Colombia	6	India	9	Pakistan				

3.1 BANGLADESH

Md. Shoriful Alam Mondal is a Ph.D. researcher at the Tokyo University in Tokyo, Japan. He previously worked as a National Team Leader with the JICA Solid Waste Management Project Team in Bangladesh. He completed his M.Sc. in Environmental Engineering at the Bangladesh University of Engineering and Technology in 2011.

3.1.1 Introduction

Bangladesh is in the grasp of environmental issues of solid waste management due to the increasing growth of the urban population in its cities (Abedin and Jahiruddin 2015). BBS (2011) reports the population of Bangladesh to be approximately 144 043 697, which includes 76.7% of the inhabitants from rural areas and the remaining from urban areas. Dhaka's population has been projected to increase rapidly in the future, thereby raising the amount of solid waste generation commensurately. A significant amount of municipal solid waste is produced daily from the six major cities of Bangladesh (Dhaka, Chittagong, Khulna, Rajshahi, Barisal, and Sylhet.). The capital city, Dhaka, contributes the most (69%) to the total waste stream (Alamgir and Ahsan 2007).

According to Waste Concern (2014), Bangladesh generates approximately 8 646 120 tons of waste annually and individuals generate 204.4 kg (450.6 lb) waste per capita annually, depending on their income level. Factors contributing to composition of the generated waste within the country are population density, lifestyles, economic conditions, fruit seasons, climate, recycling, and waste management programs (Abedin and Jahiruddin 2015). Nearly 70 to 80% of the generated waste contains organic content (Alamgir and Ahsan 2007).

3.1.2 Collection

The solid waste management system in Bangladesh is not well regulated. Three categories of management systems are practiced throughout the country: a formal system, a community initiative, and an informal system. The formal system is based on the traditional system of collection (Figure 3.1), and the transportation and disposal of waste are carried out by municipalities or city corporations of the respective areas. Collection efficiency by the formal system is between 40 and 50%. No source separation has been formalized by any city corporation. The segregation of recyclable waste with economic value, such as newspapers, bottles, cans, glass, plastics, metal, and rubber is segregated informally at the source and sold to buyers (DOE 2004).

Figure 3.1 A nonmotorized van (primary collection vehicle) carrying household waste to secondary dust bins.

According to DoE (2004), at least 80% of the generated waste can be recycled if source separation practice is employed. Three types of vehicles are used for the secondary waste collection throughout the country: conventional open trucks, demountable containers, and tractors and trailers (DOE 2004).

The collection of solid wastes is divided into two phases: primary collection and secondary collection. Primary collection consists of either direct disposal of the household waste by the owner into the dustbin or by door-to-door collection of household solid wastes. These wastes are carried to secondary dustbins from where they are further moved to the dumping site.

3.1.3 Processing and Recycling

No formal recycling system has been developed by city corporations for households, so the recycling of urban waste in the country is mainly dominated by the informal sector. Waste is recycled at different stages and involves the waste generation, storage, collection, and disposal stages (Figure 3.2). Waste pickers separate the mixed waste collected from the households and sell the recyclable waste in local markets. Nearly 10–15% of the generated waste is recycled by the informal system (DOE 2004). Since approximately 70–80% of the generated waste is composed of organic matter, local government bodies are currently replicating waste concern's model of community-based composting in a number of cities. The programmatic CDM project is implementing composting plants, using municipal organic waste from towns that include city corporations or municipalities in Bangladesh.

Figure 3.2 Transfer of solid waste from primary collection vehicle to secondary transfer stations.

(a)

(b)

Figure 3.3 (a, b) Scavengers sorting different types of waste and collecting useful materials before they are covered.

Approximately 15% of the daily generated waste is recycled (Enayetullah and Hashmi 2006). Recycling occurs in three different ways in Dhaka (Ahsan et al. 2014). Waste generators separate valuables from waste at the source and sell them to hawkers who resell them to recycle shops. Human scavengers collect (Figure 3.3) the less valuable recyclables from secondary disposal sites (SDS) and least-valued recyclables from ultimate disposal sites (UDS) and sell them to recycle shops (Ahsan et al. 2014). The recycle shops clean the recyclables and subsequently supply them to consumers or to industry as raw materials (Hai and Ali 2005).

Figure 3.4 Open dumpsite in Dhaka, Bangladesh.

3.1.4 Final Disposal

Open and unsanitary crude dumping (Figure 3.4), which is the most unhygienic system, is commonly adopted throughout the country, and nearly 35–50% of the generated waste is dumped illegally (DOE 2004). The landfill is the ultimate disposal mechanism of solid waste in Dhaka city, where only the final residuals of solid waste after recycling, re-use, or reduction are deposited. Two landfills are in Dhaka city, the Matuail landfill and the Aminbazar landfill. Wastes generated from the Dhaka South City Corporation area and the Dhaka North City Corporation area are deposited in the Matuail landfill and Aminbazar landfill, respectively. The Matuail landfill is semi-aerobic and is the only sanitary landfill in Bangladesh (BIGD 2015).

3.1.5 Major Problems

The major problems of solid waste management in Bangladesh are listed later.

- Due to lack of motivation and awareness, improper selection of technology, and inadequate financial support, a considerable portion of wastes (40–60%) are not properly stored, collected, or disposed of in the places designated for ultimate disposal (Ahsan et al. 2005). As a result, the solid waste creates environmental problems (Figure 3.5)
- In low-income countries such as Bangladesh, NGOs take the lead in composting the organic portion; nevertheless, much of the organic waste, as well as other value-less waste, remains a major problem. This often constitutes more than half by weight of the total MSW generated and requires costly removal and disposal (Ali and Harper 2004).

Figure 3.5 (a) People burning waste haphazardly in Dhaka. (b) Smoke of burned waste creating breathing problem for pedestrian.

- The absence of a national policy that encourages recycling practices and the lack of proper handling rules and standards create barriers to proper solid waste management (Abedin and Jahiruddin 2015).

- The lack of financial resources and inefficient tax collection hinder proper solid waste management.

- In a densely overpopulated country such as Bangladesh, the shortage of suitable land for the final disposal of solid waste is one of the major problems.

- There is a lack of partnership between the public sectors, private sectors, and community groups for solid waste management.

The following measures should be considered to fix the solid waste management scenario in Bangladesh:

- Source separation should be introduced and formalized at the household level for sustainable waste management.

- Legalities should be followed in opening a landfill to replace open dumpsites, and the in connection with the existing landfill site, DCC should comply with the Environment Conservation Act and Rules and Preservation Act (Abedin and Jahiruddin 2015).

- If open dumpsites are discarded, then it should be kept in mind that the government needs to provide a sustainable income source for scavengers and waste pickers.
- Composting and anaerobic digestion should be encouraged for organic waste.
- Microenterprises in waste recovery and recycling should be developed and established, and the NGOs and media should be involved in an environmental awareness program (Abedin and Jahiruddin 2015).
- Public awareness should be raised through mass media to urge the city dwellers' cooperation with proper solid waste management.

Summary of waste management in Bangladesh is presented in Table 3.1

Table 3.1 Waste Management in Bangladesh.

Population	Country (Bangladesh)	Capital (Dhaka)
Total population	144 043 697 (BBS 2011)	12 043 977 (BBS 2011)
Urban population (%)	23.30% (BBS 2011)	77.36% (BBS 2011)
Rural population (%)	76.70% (BBS 2011)	22.64% (BBS 2011)
Waste generation		
Waste generation/year	8 646 120 tons (Waste Concern 2014)	2 156 785 (JICA 2018)
Waste generation/ person/year	204.4 kg (450.6 lb) (Waste Concern 2014)	153.3 kg (337.9 lb) (JICA 2018)
% Organic waste	74.4% (Alamgir and Ahsan 2007)	70–80% (JICA 2018)
Collection – % Collected	40–50% (By formal system) (DOE 2004)	89% (JICA 2018)
Type of vehicle and collection frequency	Open truck collection, demountable container, and tractors and trailers (DOE 2004)	Open truck, dump truck, compactor (3 trips/day), container carrier (4 trips/ day) and arm roll (5 trips/ day) (JICA 2018)
% Mixed waste or % Source separated Processing	Data not available	In-house waste separation (45%) (BIGD 2015)
Material recovery facility and recycling (%)	10–15% (Recycling by informal system) (DOE 2004)	9% (JICA 2018)

Table 3.1 (Continued)

Population	Country (Bangladesh)	Capital (Dhaka)
Composting	Composting plants implemented programmatic CDM project using municipal organic waste of towns in Bangladesh (CDM n.d.)	5 community-based composting plants (one 10-12 tons/day capacity; two 3 tons/day capacity; and two 1 ton/day capacity plants) (C40 Cities 2016)
Disposal – Open dump	35–50% (DOE 2004)	Amin Bazaar is an open dumping landfill.
Disposal – Engineered/ Sanitary landfill	Only Dhaka city has one	Matuail landfill is a sanitary landfill.
Landfill gas – Collection	Data not available	Leachate collection pipes and gas vent pipes have been installed in the Matuail landfill, but gas is not collected due to poor maintenance.
Landfill gas – Flaring or LFG	Data not available	Not implemented
Waste-to-energy	Feasibility studies of Mymensingh, Cox's Bazar, Sirajganj, Dinajpur, Jessore, and Habiganj municipalities for installing a WTE facility (SREDA 2015)	Implementing the first ever waste-based power plant project at Keraniganj
Major problems	• Environmental and health problems due to open dumpsites	
	• Lack of recycling practice	
	• Absence of source separation	
	• Lack of attention to huge amount of organic waste	
Future needs	• Formalization of source separation	
	• Engineered landfill with gas collection system and MRF	
	• Composting and anaerobic digestion of organic waste	
	• Public awareness	

3.2 BRAZIL

Thiago Villas Bôas Zanon, *Solví Participações S.A., Technical Supervisor, Bachelor's degree, and M.Sc. in Geotechnical Engineering at Escola Politécnica of the University of São Paulo (USP), Brazil is a Professional Engineer in Brazil. He has been involved for more than 13 years in landfills and waste treatment site selection, studies, design, licensing, works, auditing, and operation in dozens of facilities in Argentina, Bolivia, Brazil, and Peru. He is also an instructor and coordinator for landfill in-company training for Solví, postgraduate, and specialization courses. He is a board member of ABLP (Brazilian Association of Solid Waste and Public Cleaning) and member of ISWA (International Solid Waste Association) Working Group on Landfill. He has written several national and international technical and scientific articles and publications.*

3.2.1 Introduction

In 2018, the Brazilian population was estimated to be 208 million, which was distributed among 5570 cities, 85% of whose residents live in urban areas (Azevedo, 2021; SNIS 2017). Almost ten years after the establishment of a national waste policy by the National Information System on Solid Waste Management ("SINIR"), there remains a great lack of data in Brazil. It also explains the lack of updated data. In 2017, the national MSW generation was estimated to be 78.4 million tons, resulting in per capita generation of 376 kg/person/year (829 lb/person/year). The gravimetric composition of the waste, estimated in 2008, was 51% organics, 14% plastics, 13% paper, 3% metals, 2% glass, and 17% of other materials.

In 2018, the population of São Paulo, the most populous city in Brazil and one of the most populous of the world, was estimated to be 12.2 million, 99% of whom lived in urban areas (IBGE 2018). Despite being one of the richest cities, having a municipality with resources, in addition to having a waste management system that is considered of above average quality compared to the rest of the country, there is lack of some specific data along with some updated data. In 2017, the MSW generation was estimated to be 3.9 million tons, which translates to per capita generation of 320 kg/person/yr (705 lb/person/year), which is lower than the amount previously presented in Brazil (376 kg/person/yr). It demonstrates a possible inconsistency of the data, considering that the population of the city of São Paulo has a significantly higher income population than the Brazilian average. The gravimetric composition of waste, in 2012, was 51% organics, 32% recyclable materials, and 17% other materials.

3.2.2 Collection

The household collection index in Brazil is 91%, and almost all (98.4%) of the collected waste is mixed and is not source-separated. The typical trash collection frequency is every two days, while recyclables are collected once per week. Rear loaders are the typical vehicle used for both types of collection (mixed or source-separated) (Figure 3.6)

The household collection index in São Paulo is 100%, higher than the national average, as expected (PGIRS 2014). Almost all the collected waste is mixed (97.7% of the total MSW collected), without source separation, which is close to the national average. The mixed material is collected daily in the downtown area and every two days in the rest of the city. In areas where recyclables are collected, it is done on a weekly basis. Rear loaders are the typical vehicle used for both types of collection (mixed or source separated).

3.2.3 Processing and Recycling

The MSW is not composted on a large scale, although official data points to a small 0.8% share of MSW being sent for composting in 2008 (PNSR 2012). There are many

Figure 3.6 Collection vehicles and workers.

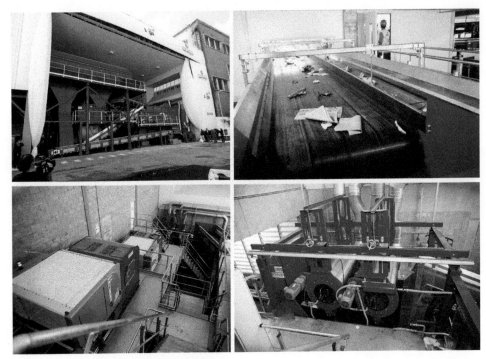

Figure 3.7 Material recovery facility.

material recovery facilities (MRFs) dispersed around the country (Figure 3.7), but only five of them are automatic or semi-automatic. Most of the facilities use manual sorting that is done by scavengers' cooperatives that lack sufficient organization and efficiency and are usually subsidized with public resources. At the end of 2018, 5 composting yards with processing capacity of approximately 3000 tons per year were installed for the processing of organic residues (fruits, vegetables, and greens).

3.2.4 Final Disposal

In 2017, 41% of Brazil's MSW was disposed of in open dumps, while 59% was disposed of in engineered and sanitary landfills. The size of the problem can still get worse if one considers that at least 56% of the cities still use dumps.

Due to the existence of large and populous urban conglomerates and regions, there are several sanitary landfills that receive practically all of the MSW generated, which represents a very large amount of waste. Moreover, the favorable conditions in the landfills (mainly organic content in the residue and humidity) favor the use of active landfill gas (LFG) collection (Figure 3.8) and burn on gas power plant, totaling

Figure 3.8 Biogas collection.

17 facilities currently in Brazil. As there are no legal or normative obligations for the efficient collection and destruction of LFG methane (except in São Paulo, which requires the environmental licensing of new landfills or unlicensed expansions of existing landfills), landfills that have an efficient collection process and flare for centralized biogas are associated with old, certificated reduction emissions (CERs) associated with the Kyoto Protocol or voluntary initiatives of some high-income economies.

For decades, all of the MSW generated in São Paulo city has been disposed of in engineered and sanitary landfills. Currently, there are two landfills that are closed but still have active biogas collection and burning on gas power plant for power generation. Besides this, there are three landfills in the city and nearby that receive MSW, and two of them are located where there are also gas power plants for power generation.

3.2.5 Major Problems

The major solid waste management problems in Brazil are as follows:

- The Brazilian Constitution establishes that MSW management is the responsibility of the local government; however, the taxes are collected at the federal and state levels. Few resources are available at the local level, a situation that is aggravated by the crisis of economic growth that Brazil has been facing since 2014. The major problem, currently, is the economic capacity of the municipality to ensure that the resources required for minimally adequate waste management are available.

- The public managers need to have improved technical capacity.

- The landfills that receive the MSW from São Paulo have a significant remaining useful life, but when they are closed, the new ones will be put in areas that are farther away. The associated costs for the final disposal, which are related to the cost of transporting the waste, will then increase.

- Since the city does not have a specific tax collection system for the generators, it is difficult to find sustainable solutions in waste management, as it would require the expenditure of a greater amount of economic resources.

Some solutions for fixing the above problems are already widely known and implemented globally. Everything considered, landfills are the most viable solution for Brazil, both from a technical and economic point of view.

- Brazil is a country with continental dimensions that present large vacant available areas. It needs to close its dumps and seek consortium or regionalized solutions to compensate for its lack of economic resources. It can optimize its economy by grouping cities together, through the implementation of regional landfills, like the United States did in the 1980s, when it eradicated 20 000 dumps in less than two decades. In addition, Brazil should search logistically for the best sites for the collection bases, transfer stations, and sanitary landfills to reduce transport costs.

- A specific tax should be collected from the generators to guarantee the necessary resources for the management and execution of an adequate waste management system. This represents a strategy that has not been often used in the country but could be instrumental in solving some of their problems.

- Economies of scale and logistics should be considered as solutions that seek to enhance the generation of resources and energy from waste, and these actions should be intensified after eliminating the environmental problems from the closure of the dumps.

- Technology should be used to reduce the amount of waste that is landfilled (both organic and recyclable). Source separation and collection should be enforced for sustainable waste management.

- An environmental education program is essential as a venue for sharing knowledge about the proper way to separate waste at the source and reduce the overall amount of waste generation.

- Waste mining and technologies for waste recovery should be evaluated. Since there are several closed landfills in São Paulo, the mining of such landfills can be a viable solution.

The summary of Brazil's waste collection system is presented in Table 3.2

Table 3.2 Waste Management in Brazil.

Population	Country (Brazil)	Major City (São Paulo)
Total population	208 494 900	12 176 866
Urban population (%)	85	99
Rural population (%)	15	1
Waste generation		
Waste generation/Year	78.4 million tons (172 billion lb)	3.9 million tons (8.6 billion lb)
Waste generation/ Person /Year	378 kg (833 lb)	320 kg (705 lb)
% organic waste	51	51
Collection - % Collected	91	100
Type of vehicle and collection frequency	Rear loaders Frequency: every 2 d	
% mixed waste or % source separated	Mixed waste: 98.4 Source separated: 1.6	Mixed waste: 97.7 Source separated: 2.3
Processing		
Material recovery facility and recycling (%)	–	–
Composting	0.8%	0.5%
Disposal – Open dump	41%	0%
Disposal – Engineered/ Sanitary landfill	59%	100%
Landfill gas – Collection	17 facilities	3 facilities
Landfill gas – Flaring or LFG	17 facilities	3 facilities
Waste to energy	0%	0%
Major problems	• Lack of economic resources • Lack of technical capacity of public managers	• Lack of economic resources • Lack of areas for new landfills in the future
Future needs	• Consortium or regionalized solutions • Adopt a specific tax to be collected directly from the generators	• Source separation • Environmental education program • Waste mining of closed landfills

3.3 COLOMBIA

Vishwas Vidyaranya is the cofounder of Ambire, a company focused on consulting and digital solutions in circular economy and climate change. He has over 10 years of international experience in circular economy, waste and water management, financial analysis, climate change resilience, and sustainable cities. He holds a Master's degree in Sanitary Engineering and Waste Management from Leibniz Universität Hannover, Germany, and is based out of Bogotá, Colombia. Vishwas has implemented over 4000 tons per annum capacity of decentralized waste recycling systems and designed over 150 000 tonnes per annum of centralized waste treatment systems. This includes five material recovery facilities ranging from 20 TPD to 250 TPD capacities.

He is actively involved in professional organizations like ISWA and its Young Professionals Group (YPG). He is also a member of ISWA's working group on landfills. He has several publications and a US Patent on the energy management system to his credit.

3.3.1 Introduction

Colombia is one of the fastest growing economies of Latin America with a population of approximately 49 million (Banco Mundial 2019) and a GDP of about 315.595 million USD. The country is divided into 33 states and 1102 municipalities, with Bogotá as its capital. The per capita waste generation varies widely among different states and is estimated between 0.64 kg/d (1.41 lb/d) and about 1 kg/day (2.2 lb/d) (SSPD 2017; UN Environment 2019). A detailed analysis of countrywide data by the public authority at the end of 2017 indicated a generation of 30.081 tons/day (66 317 lb/d) with a YoY increase of about 2% (SSPD 2017). Over 53% of the total waste is generated in three states: Antioquia, Valle del Cauca, and Bogotá D.C (SSPD 2017). Most of the waste is organic (52–60%), followed by plastics (11–13%) and paper and cardboard (10–16%) (Sarmiento 2017; UN Environment 2019).

3.3.2 Collection

Colombia is an upper-middle income country and has a coverage of waste collection of about 82%. In some cities, such as Bogotá, the coverage is over 95% (UN Environment 2019). Colombia has enforced segregation of waste at the source, and the category for separation depends on the waste management plans laid down by the municipalities. Bogotá, for example, mandates separating recyclables from the rest. The legislation "Decreto 1077 de 2015" describes the ground rules for separation of source, separate collection of waste, vehicle standards, quantification, transfer stations, and final disposal (Ministerio de Vivienda, Ciudad y Territorio, Colombia 2015).

The law requires that the vehicles transporting organic fractions should be in closed vehicles to avoid odor and vectors. For nonrecyclables, the collection frequency is at least twice weekly. The recyclables are required to be transported in a manner that does not deteriorate their characteristics. As in many developing countries, the informal sector is very active in Colombia, and they are responsible for separating the recyclables and recycling them. In 2016, the government introduced a landmark legislation, "Decreto 596 de 2016," aimed at integrating the informal sector into the formal waste management system. This legislation gives a window for informal recyclers to register and formalize their organizations with the public authorities. It also gives them the right to collect the recyclable waste; the utility company oversees collecting the nonrecyclable waste. The recyclables are transported by these recyclers to a transit point by using hand carts, tricycles, or motorized transports. A portion of the waste tariff collected by the municipality, or the utility companies, is redistributed to the recycler organizations. Through this legislation, the informal sector also gets trained and enabled to improve their management and operations. This helps in improving their working conditions and in increasing the recycling rates.

3.3.3 Processing and Recycling

The recycling rates are estimated to be over 17% in the country (Sarmiento 2017). This can be attributed to the activity of the informal sector and formalized recycler associations (Figure 3.9). These organizations have small capacities. More than 50% of the registered recyclers collect less than 100 tons (220 000 lb) of recyclable materials per month (Superintendente de Servicios Públicos Domiciliarios 2018), and more than 40% of the organizations have fewer than 50 recyclers who collect from various

(a)   (b)   (c)

Figure 3.9 (a) PET recycling facility, (b) informal recycling, (c) nonrecyclable and recyclable containers.

locations such as residences, industries, and commercial centers. Only about 40 tons/day or 0.1% of the total waste collected by the utility companies is processed in treatment plants, which is almost negligible.

3.3.4 Final Disposal

The common methods of disposal are sanitary landfills (approximately 97%) followed by open dumpsites, transition, or temporary cells, and burying (SSPD 2017). About 78 % of the municipalities dispose of their waste in authorized disposal sites and the remaining use either open dumps or no data are recorded (SSPD 2017).

Doña Juana, Bogotá's landfill, began with a project of LFG capture to reduce emissions (Figure 3.10). In 2009, the landfill registered with the UNFCC under the Kyoto Protocol project listings and successfully implemented the project in 2016 when the facility started generating approximately 1.7 MW of electricity. The capture and burning of methane have helped reduce approximately 800 000 tons/year of CO_2 emissions (Biogas Doña Juana 2019). This is an exemplary pilot project for reducing emissions from final disposal sites in Colombia, but the landfill operations and management are still highly inadequate across the country. Doña Juana in particular has experienced many problems related to health and the environment.

3.3.5 Major Problems

The major problems of solid waste management in Colombia are described later.

- Open dumping and illegal disposal are the biggest challenges for final disposal in several municipalities across Colombia (Figure 3.11). Such methods can have high impacts on public health and can lead to indirect costs of up to 10 times the costs of effective waste management (UNEP 2015).

Figure 3.10 Bogotá Landfill.

Figure 3.11 Quibdó Dumpsite.

- The short remaining life span of several authorized landfill sites is a major concern. About 40% of the landfills across the country have an estimated life span or capacity to accept waste of less than 3 years. The biggest landfill of the country, Doña Juana, which receives approximately 7000 tons/d (15.4 million lb) of solid waste from Bogotá has had critical operational issues over the last few years. In 2015, the landfill had a major slip failure partially due to leachate build-up and improper management, which resulted in over 370 000 tons of waste being spilled from the site (Bnamericas 2015; SSPD 2017). This resulted in an emergency that led to severe health impacts in the neighborhood and resulted in severe odor, vectors, and other issues. In the last ten years, similar problems have occurred in other major landfills such as Bucaramanga and Boyaca.

- The waste management tariffs have been increasing significantly (in some cases up to 30%) to manage the high costs of collection and disposal (El Tiempo 2018).

- The entire system of waste management is centralized, which results in high financial stress for the municipalities that have budget constraints.

- The legislation in Colombia seems to be exhaustive in defining the solid waste management systems. Although the informal sector has been integrated into mainstream waste management and increased the recycling rates in the country, effective waste management is still a big concern.

- Segregation at the source is not significant enough to support the recycling system. In several municipalities, the collection infrastructure is not equipped for separate collections of waste.

- Even though the official reports suggest that 78% of all the of the municipalities, accounting for more than 97% of the solid waste, have access to authorized landfill sites (SSPD 2017), the lack of treatment facilities is a big problem. Several landfill sites have had issues of poor operation and mismanagement.

The following measures should be considered for fixing the solid waste management problems in the cities of Colombia.

- Their waste has a high (60%) composition of organic material, which has the potential for producing compost and biogases. It is not currently being utilized in this way, but with proper policy changes, it could not only reduce the burden on final disposal but also reduce the emissions and generate by-products such as compost or biogas.
- Colombia is a growing agricultural economy and generating compost could help transition towards a circular economy.
- Decentralized waste management must be encouraged, at least for the bulk producers such as commercial centers, hotels, and large campuses. This will reduce the burden on municipalities, both financially and in terms of resources.
- Stronger policies aimed at organic waste treatment and a circular economy are needed to improve the waste management situation in the country.

The summary of waste management system of Colobia is presented in Table 3.3

Table 3.3 Waste Management in Colombia.

Population	Country (Colombia)	Capital (Bogotá)
Total population	49 million	8 million
Urban population (%)	80	>99
Rural population (%)	20	<1
Waste generation		
Waste generation/Year	11.5 million tons (25.4 billion lb)	2.2 million tons (4.85 billion lb)
Waste generation/ Person/Year	233.6–310 kg (515–683 lb)	285 kg (628.3 lb)
% organic waste	60	60
Collection – % collected		
Type of vehicle and collection frequency	Frequency: at least twice per week for nonrecyclables	Compactor trucks Frequency: thrice per week

Table 3.3 (Continued)

Population	Country (Colombia)	Capital (Bogotá)
% mixed waste or % source separated	Mostly mixed	Separate containers and collections are being implemented but is mostly mixed
Processing		
Material recovery facility and recycling (%)	0.1%	No
Composting	–	No
Disposal – Open dump	Yes	No
Disposal – Engineered/ Sanitary landfill	97%	Yes
Landfill gas – Collection	–	Yes (partial)
Landfill gas – Flaring or LFG	–	Yes
Waste to energy	No	–
Major problems	• Open/Illegal dumping and no treatment facilities	
	• Bad operations of landfills	
	• High tariffs and ineffective centralized waste management	
	• Low segregation at source	
Future needs	• Treatment facilities for compost and biogas production	
	• Decentralized waste management	

3.4 ETHIOPIA

Eshetu Assefa has a master's in Water Supply and Environmental Engineering. He is a lecturer on the Faculty of Civil and Water Resources Engineering at Bahir Dar Technology Institute, Bahir Dar University, Ethiopia. He has over 13 years of teaching, design, research, and practical experience in the areas of water supply and environmental engineering. In research, he has experience in solid waste management, adapting and developing wastewater treatment technologies like the use of urine for agriculture, cocomposting and gray water reuse, urban sanitation, and water quality and supply. He was

a chair holder of Water Supply and Sanitary Engineering at Bahir Dar Technology Institute for more than three years. He was also coordinator of the Ethiopian partners in CLARA (Capacity Linked Water Supply and Sanitation Improvement for African's Peri-Urban and Rural Areas), a project funded by the European Union under the Fp7 framework funding program.

3.4.1 Introduction

The population of Ethiopia currently is estimated to be 114 963 588, of which 21.3% live in urban areas and 78.7% live in rural areas, according to UN data. According to the 2007 Census conducted by the Ethiopian Central Statistics Agency, there were 2.7 million people living within the jurisdiction of the capital city of Ethiopia, Addis Ababa, (Federal Democratic Republic of Ethiopia Population Census Commission 2008). The Climate and Clean Air Coalition (CCAC) MSW Initiative City Assessment prepared by the US EPA and the Addis Ababa City Government Solid Waste Disposal and Recycling Project Office (CCAC MSW Initiative 2014) indicates that the population of the city in 2014 was 3.9 million. Currently, Bahir Dar City has a population of 418 310 of which 79.3% live in the urban areas and 20.7% live in the peri-urban and rural areas.

According to the US EPA report in 2006, the per capita amount of solid waste generated in Ethiopia ranged from 102.2 to 303 kg/year (225.3–668 lb/yr). The Addis Ababa Solid Waste Disposal and Recycling Project Office estimates that the city generates 730 000 metric tons of MSW annually (CCAC MSW Initiative 2014), and that the per-capita waste generation rate is approximately 164.3 kg per year (362.2 lb/yr) (CCAC MSW Initiative 2014). Based on the income level of households in 2019 in Addis Ababa, the rate of generation of waste ranged from 219 to 767 kg/person/yr (Tassie et al. 2019).

The residential per capita waste generation rate in the city of Bahir Dar in 2014 was 104 kg/year (229 lb/yr) and from all waste sources was 194 kg/person/yr (427 lb/person/year). According to the socioeconomic survey conducted for structural preparation of the city in 2019, the per capita waste generation rate of the residential area was 124.5 kg/yr (274.5 lb/yr), and the annual waste generation of the city from residential area was 51 452 metric tons (0.11 billion lb).

The average composition of solid waste generated in the city of Addis Ababa and Bahir Dar is presented in Tables 3.4 and 3.5, respectively. As per the results of the study, 65% of the Addis Ababa City MSW is organic and 86.6% of Bahir Dar City MSW is food waste.

3.4.2 Collection

According to a study conducted in Ethiopia in 2011, no formal waste segregation was done at any stage of the waste collection chain in any of the municipalities (Community Development Research 2011). Even currently, most of the municipalities do not sort the waste at the point of generation. Everything generated is mixed at the point

Table 3.4 MSW composition in Bahir Dar City.

Component	Percentage
Food	86.6%
Paper	3.3 %
Plastic, leather	2.2 %
Glass	0.6 %
Textile	2.2 %
Metals	0.3 %
Others	4.8%
Total	**100%**

Source: Based on UNEP (2010a).

Table 3.5 MSW composition in the City of Addis Ababa.

Component	Percentage
Organic	65%
Paper	3%
Plastic	3%
Tires	3%
Metal	2%
Glass	1%
Others	23%
Total	**100%**

Source: Data from CCAC MSW Initiative (2014).

of generation, collected, and transported to the open dumpsite except for a very few households that separate their organic waste from the rest for the purpose of producing compost for their gardens.

In Addis Ababa, the primary collection is done door-to-door by micro-and-small-scale enterprises (MSEs). The MSEs move the collected waste with hand carts to "skip points" and deposit it in 8 m³ containers. The secondary waste collection from skip points to the "Repi" open dumpsite is the responsibility of the subcity cleansing authority offices. The municipality's trucks retrieve waste from the skip points and transport it to the city's open dumpsite in the Repi neighborhood. According to the Artelia, the average emptying frequency of each skip container is twice a week (Artelia 2013). The Addis Ababa City municipality has waste collection vehicles that are equipped with internal compactors that are mainly used to enhance the efficiency of the waste collection. The Solid Waste Disposal and Recycling Project Office

(a)   (b)

Figure 3.12 (a) Primary waste collection from households to skip points.
(b) Loading of collected waste from skip points.

(a)   (b)

Figure 3.13 (a) Secondary waste collection vehicle at the working face of dumpsite and dumpsite workers. (b) Secondary waste collection by the Municipality, Bahir Dar city.

estimates that Addis Ababa City has a collection efficiency of nearly 80% (CCAC MSW Initiative 2014).

In Bahir Dar city, the municipality collects solid waste by employing five MSEs and one private company, Dream Light PLC. The five enterprises use hand carts to collect waste door-to-door, which they deposit in skip points, where the municipality truck (Figure 3.12b) picks it up and takes it to an open dumpsite that is located 3–4 km (1.9–2.5 miles) from the center of the city. The private company uses hand carts for the door-to-door collection and 9 trucks for the secondary collection (Figure 3.12a, b, Figure 3.13) and trip to the open dumpsite. Based on a socioeconomic survey of more than 1000 households in 2019, the waste collection frequency in different localities in Bahir Dar City is once in a week (44%), every 15 days, every 21 days, once a month, or not at all. The waste collection efficiency of Bahir Dar city was 71% in 2010, according to the UNEP (2010b) study. Based on the socioeconomic survey conducted for the purpose of preparing a city structural plan, the collection efficiency of the city is about 53%.

The efficiency of waste collection in urban areas of Ethiopia in 2011 ranged from 30 to 82% (Community Development Research 2011).

3.4.3 Processing and Recycling

The practice of waste composting in Ethiopia is initiated at the household level, by solid waste collection cooperatives, small and micro enterprises, private companies (Figure 3.14b, c), and city municipalities, depending on the extent of compost utilization.

Based on a report by the CCAC Municipal Solid Waste (MSW) Initiative in 2014, approximately 5% of all of the waste that is generated in Addis Ababa is recycled (CCAC MSW Initiative 2014). Informal recycling plays a key role in recycling iron, other metals, and plastic bottles, although it is carried out in an unorganized fashion. There are also many individual entrepreneurs who walk the streets calling out for household recyclables. According to the CCAC MSW Initiative report in 2015, there are nearly 1000 recyclers who pick plastics, iron, and other materials from the Repi ("Koshe") dumpsite.

(a) (b)

(c) (d)

Figure 3.14 (a) Refuse drived fuel, Bahir Dar city, (b) compost produced by private company, (c) composting by the municipality, (d) collected plastic bottles for recycling.

The recycling of waste in Ethiopia also attracts the attention of the business sector. Currently, there are around 10 to 40 recyclers who process mixed thermoplastics, 10 plastic bottle recycling companies, 2 paper recyclers, 7 aluminum recycling companies, and many steel recycling companies (Global Business Network Program 2020).

In the city of Bahir Dar, the capacity of the municipal composting plant is small and uses only 0.5 metric ton per day (1102 lb/day) of organic solid waste to produce compost (Christian 2012). In 2010, there was less than 1% recycling of valuable waste materials, which was mainly done at the open dumpsite by the informal recyclers. Plastic bottles are recycled by informal recyclers both before and after collection of the waste (Figure 3.14d). The Dream Light PLC produces high-quality charcoal briquettes from waste (Figure 3.14a). The city municipality and Dream Light PLC produce compost from bio-solid waste that is collected mainly from market centers and hotels and sold for urban agriculture to nearby farmers (Figure 3.14b, c).

Ethiopia does not have any kind of a material recovery facility (MRF) to sort and reuse recyclable waste materials.

3.4.4 Final Disposal

None of the municipalities in Ethiopia except Dire Dawa, Mekele, and Hawassa have an engineered landfill into which they can dispose of collected waste. Instead, the waste is dumped in an open area close to or within the city limits (Community Development Research 2011). Addis Ababa city has one dumpsite at "Repi" ("Koshe"), the largest and oldest open dumpsite in Ethiopia, which has been in use for more than 50 years (Community Development Research 2011; CCAC MSW Initiative 2014). In 2013, the dumpsite received approximately 4.5 million m³ of waste (Artelia 2013). In 2018, it was transformed into a new waste-to-energy plant that incinerates up to 1400 metric tons (3 million lb) (80% of the city's waste) per day and supplies the city of Addis Ababa with 25% of its household electricity needs (UN Environment Program 2019).

The construction of the first waste disposal cell of the engineered landfill at "Sendafa" was completed in 2015. It is 120-hectare in size and is located approximately 25 km (15.5 miles) from the Addis Ababa city center. The new sanitary landfill was constructed primarily for the purpose of closing the existing open dumpsite and disposing of waste in a managed and engineered way. After only three months of operation, the engineered landfill at "Sendafa" was not functioning due to social, economic, and political reasons. As a result, waste from the city was disposed of once again in the "Koshe" open dumpsite. In 2017, the landfill failed, mainly due to sliding caused by its operating over capacity, and 113 people who were living in and around the landfill lost their lives.

Bahir Dar city only has one open dumping site for disposal of the collected waste from the city. The municipality does have two excavators and one bulldozer that they used to distribute and compact the waste (Figure 3.15a). The city recently introduced a new landfill disposal technique for managing the dumped solid waste at the open

(a) (b)

Figure 3.15 (a) Equipment used in the working face of dumpsites. (b) Installation of pipes for the FUKUOKA method.

dumpsite. The technique is said to be a "FUKUOKA" method, which is a semi-aerobic landfill disposal technology. Perforated pipes (Figure 3.15b) are installed at the bottom of the landfill to create conditions in which air is easily introduced to the bottom of the landfill layer and enables quick drainage of the leachate.

3.4.5 Major Problems

Ethiopia's major waste management problems are listed later.

- **Improper waste management systems and absence of sanitary landfills:** Mainly due to a lack of resources, including budget, capacity, technology, and strict environmental laws and policies.
- **Scarcity of land:** Very high land lease values in almost all of the cities in Ethiopia create a scarcity of land for final disposal.
- **Inefficient waste-to-energy facility:** The waste collected from Addis Ababa city and directed to the waste-to-energy facility at "Koshe" is mixed, and the largest proportion is composed of food waste, which has very high moisture content. This makes the facility inefficient for producing energy.
- **Environmental pollution and health problems:** Open dumping of the collected waste and dumping of uncollected waste into the environment in the urban areas of the country negatively impact the environment components like surface and ground water sources, soil, and air (Abebe and legese Abitew 2018). This results in the creation of human health risks, either directly or through the food chain.

The following measures can be undertaken to fix the problems discussed earlier.

- Reliable data on the characterization and quantification of municipal solid waste should be recorded by the municipalities, as it is an important guideline for planning and designing waste management facilities, resource recovery techniques, and waste-to-energy systems.
- Separation of waste at the point of generation should be planned and in place to be effective at each step of the solid waste management chain, including resource recovery and the waste-to-energy process.

- City municipalities should work to improve waste collection efficiency throughout the city, as it is the main source of revenue for a better waste management service.

- The government and city municipalities should work together to improve public attitudes and enhance public awareness and participation, as it potentially affects the whole solid waste management system: household waste storage, waste segregation, recycling, collection frequency, willingness to pay for waste management services, and opposition to the siting of waste treatment and disposal facilities.

- City municipalities should work to organize the waste pickers and incorporate them into the waste recycling system, as it is essential to create a market chain for the recycled materials that could change the waste management system.

- The government should work to increase the capacity of institutions to update existing environmental policies, strategies, and regulatory frameworks, as well as to establish strict enforcement mechanisms on waste management service providers and waste producers. This will help the city municipalities maintain the quality of the environment components and reduce the number and severity of human health risks.

- The municipality needs to commit to getting the needed financial resources and providing training for their staff, to enhance the institutional capacity to provide the required municipal infrastructure and capable human resources for an adequate solid waste management system.

In summary, if the challenges of building a solid waste management system in Ethiopia are to be met, there will need to be strong public engagement, financial management, and institutional commitment. The summary of current waste management of Ethiopia is presented in Table 3.6

Table 3.6 Waste Management in Ethiopia.

Population	Country (Ethiopia)	Capital (Addis Ababa)	Second largest city (Bahir Dar)
Total population	114 963 588	2.7 million	418 310
Urban population (%)	21.3	–	79.3
Rural population (%)	78.7	–	20.7
Waste generation			
Waste generation/Year	–	730 000 metric tons (2014)	Residential: 51 452 metric tons (2019)

Table 3.6 (Continued)

Population	Country (Ethiopia)	Capital (Addis Ababa)	Second largest city (Bahir Dar)
Waste generation/ Person/Year	102.2–303 kg (2006) (225–668 lb)	164.3 kg (362 lb) (2011) Based on Income: 219–767 kg (1690 lb) (2019)	Residential: 104 kg (229 lb) (2014) From all sources: 194 kg (428 lb) (2014) Residential: 124.5 kg (274.5 lb) (2019)
% organic waste Collection – % Collected	30–82% (2011)	65% (2014) 70% (2013) 80% (2014, 2017)	72.4% (2014) 71% (2010) 53% (2019)
Type of vehicle and collection frequency	Solid waste collection truck, hand carts, and normal trucks, which are not designed for solid waste. Frequency: once a week, every 15 d, every 21 d, once a month, and no service at all.	Solid waste collection truck, hand carts, and normal trucks, which are not designed for solid waste. Frequency: once a week, every 15 d, every 21 d, and once a month.	Hand carts and normal trucks that are not designed for solid waste collection. Frequency: once a week, every 15 d, every 21 d, a month, and no service at all.
% mixed waste or % source separated	No formal segregation done in any stage of the waste collection chain in all the municipalities		
Processing Material recovery facility and recycling (%)	Recycling of plastics, metals, paper, aluminum. But there is no MRF.	Recycling of plastics, metals, paper, aluminum. 5% recycled (2011) There is no MRF.	Recycling of plastics, metals. Less than 1% recycling in 2010. There is no MRF.
Composting	Compost from waste is produced at the household level, by solid waste collection from cooperatives, small and micro enterprises, PLC, and the municipality in a very small scale.	5% of the bio-solid produced in the city was composted in 2011.	At the household level, by small and microenterprises, private company, and the municipality at the dumpsite. 0.5 ton per day of organic waste is used to produce compost (2012).

(Continued)

Table 3.6 (Continued)

Population	Country (Ethiopia)	Capital (Addis Ababa)	Second largest city (Bahir Dar)
Disposal – Open dump	Open dumping	Open dumping	Open dumping
Disposal – Engineered/ Sanitary landfill	Available in few cities – Dire Dawa, Hawassa, and Mekele	At "Sendafa" but not functional	Not available
Landfill gas – Collection	No landfill gas collection system installed anywhere in the country except at "Repi" ("Koshe") dumpsite	At "Repi" dumpsite	Not available
Landfill gas – Flaring or LFG	Flaring	Flaring	Not available
Waste-to-energy	The "Repi" dumpsite – uses 1400 tons of waste every day to produce energy	The "Repi" dumpsite – uses 1400 tons of waste every day to produce energy	Not available
Major problems	• Lack of proper waste management system • Absence of engineered landfill • Scarcity of land • Inefficient waste-to-energy • Environmental pollution • Health problems		
Future needs	• Continuous and reliable recording of data • Source separation and resource recovery from households • Improvement of waste collection system • Strong public engagement • Financial management • Institutional commitment		

3.5 GEORGIA

Medea Chachkhiani graduated from Tbilisi State
University and started her working experience at the
Laboratory of Renewable Energies as a microbiologist.
Medea developed a deep interest in waste management
as a result of her eight years of intense scientific investigations dedicated to biomass and MSW conversion to
energy. She has an experience as a head of an Environmental Protection Unit at the Solid Waste Management
Company of Georgia under the Ministry of Regional
Development and Infrastructures. She has been involved
in several activities related to the processes of closure and
after-care of old open dumpsites and establishing high-
standard SWM systems in different regions of the country.

3.5.1 Introduction

Solid waste management is a matter of national, regional, and local concern in Georgia.
There are ten regions and four self-governing cities in Georgia: Tbilisi, the capital of
Georgia, and 63 municipalities. Georgia is inhabited by 3.7 million people, with 59.5% of
them from urban areas and 40.5% of them from rural areas. The waste generation rate
in Georgia is 275 kg per capita per day; in Tbilisi city, it is 343 kg (756 lb) per capita per
day. The waste stream in Georgia is constituted of 47% organic waste. The municipal
solid waste composition was studied on a seasonal basis in the Kakheti and Adjara
regions, the average amounts of which are given in Figure 3.16. The government of
Georgia (GoG) established a centralized Solid Waste Management Company of Georgia

Figure 3.16 Composition of
municipal solid waste in Georgia.

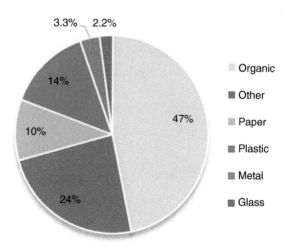

(and Infrastructure, MRDI, in SWMCG) under the Ministry of Regional Development 2012, which is mainly responsible for solving the problems of nonhazardous municipal landfills throughout the country, excluding the Adjara region and Tbilisi.

3.5.2 Collection

Source separation, which is one of the major aspects of sustainable waste management, has not yet been introduced in Georgia. Collection and transportation of wastes (Figure 3.17) throughout the country are managed by self-governance. Corresponding services are carried out by 100% state-owned limited liability companies (Ltd) or governmental agencies. The efficiency of municipal solid waste collection is much higher in big cities as compared to rural areas. To be specific, 100% waste collection is done in urban cities while in mountainous regions and villages it drops to 18%. The poor road infrastructures in some rural areas and mountainous regions make the waste collection difficult and thus considerable number of villages are left to deal with their own waste. Consequently, the residents of those regions, who are unaware of the rules of proper and sustainable waste management approaches, dump their waste such as putrescent organics, plastic bags, single-use plastic bottles, and batteries into ravines and riverbeds.

3.5.3 Processing and Recycling

As a consequence of the poor economy, MSW in Georgia contains less packaging waste. Organic waste amounts to almost half of the whole bulk of solid wastes and contains about 60% water, which decreases the quality of the secondary resources of the recyclable wastes, such as paper, cardboard, and woody materials. Accordingly, separated secondary materials are not marketable, and the price per ton of recycled wastes is too low. Source sorting of wastes needs basic changes, and the entire WM system needs to be re-equipped with relevant techniques.

Figure 3.17 Primary collection vehicle.

3.5.4 Final Disposal

There are 60 official landfills in operation throughout Georgia, but only three of them are sanitary landfills (located in Tbilisi, Rustavi (Figure 3.18), and Borjomi). The environmental parameters are controlled either through EIA (an environmental impact assessment document) or Conditional Plans approved by the Ministry of Environmental Protection and Agriculture of Georgia. Problems resulting from untreated solid municipal wastes (MWs) disposed of at existing landfills are made greater by the fact that none of the waste is sorted at the household level or at any other level. Mixed wastes from households and enterprises are collected in containers and then are disposed of in landfills where they are pressed by a hard technique and covered with a soil layer, but this does not occur regularly. Such approach partly prevents the spread of light fractions from mixed, unsorted wastes from the landfill site by the wind; however, spontaneous inflammation of methane, contained in biogas formed as a result of anaerobic fermentation of the organic fraction of municipal solid wastes, occasionally takes place. Methane ignition at landfill sites is followed by slow burning of wastes that results in the formation of dense smoke and the release of objectionable odors and toxic compounds into the atmosphere. Up to 2012, most of the official landfills, which are actually open dumps, were not even fenced (Figure 3.19). The infrastructure and special inventory, as well as techniques for operation, did not exist. At most sites, there were no lights, sources of water, or roads. Municipal

Figure 3.18 Rustavi sanitary landfill.

Figure 3.19 Nikea landfill before fencing.

clean-up service cars, animals, and people entered the landfill areas without any control or safety norms. It is noteworthy that the stabilization of hospital wastes via 14 small-scale incinerators operating all over the country was almost the only positive fact in the scope of solid waste management in Georgia.

Almost all of the 3668 villages (National Statistics Office of Georgia), excluding conflict areas, solve the waste management problems by forming uncontrolled open dumps, the number of which is assumed to exceed 3000 (rough estimation by NGOs). In rural areas, open dumpsites are primarily located close to pastures, settled areas, and small rivers; consequently, sanitation problems and environmental pollution result in numerous health problems for the local population. Harmful intermediate products of waste decomposition include increased concentrations of heavy metals in the soil; surface and groundwater pollution in the areas surrounding open dumps; easy access to unfenced areas by domestic animals; and people representing the informal sector carry a risk of spreading diseases.

3.5.5 Major Problems

The major problems of the waste management system in Georgia are listed later and proper step-by-step approaches needs to be undertaken to solve these problems.

- Old infrastructure and old WM system with minor changes.
- Lack of hazardous waste management system.
- Formation process of common database on WM on country and local levels.
- Lack of EPR and PRO (Producer Responsibility Organization) systems.

- Very low private sector and/or informal sector participation in WM initiatives.
- Lack of governmental programs to support recycling activities.
- Legacy wastes.
- Obsolete tariff system.
- Lack of source separation.

The following steps need to be taken for a sustainable approach toward waste management in Georgia:

- Reform the current WM system.
- Proper collection and disposal of hazardous waste.
- Establishment of a reliable database on the WM of the country, including local-level data
- Establishment of social initiatives to involve producers, consumers, private sectors, and informal sectors in WM.
- Considerable effort by government organizations to support and fund recycling activities and source separation.
- Create public awareness of the need for recycling and source separation.

The summary of waste management system in Georgia is presented in Table 3.7

Table 3.7 Waste Management in Georgia.

Population	Country (Georgia)	Capital city (Tbilisi)
Total population	3.7 million	1.2 million
Urban population (%)	59.5%	–
Rural population (%)	40.5%	–
Waste generation		
Waste generation/Year	1 017 800 tons (2.2 billion lb)	411 733.05 tons (0.9 billion lb)
Waste generation/ Person/Year	275 kg (606 lb)	343 kg (756 lb)
% organic waste	47%	
Collection – % Collected	80%	

(Continued)

Table 3.7 (Continued)

Population	Country (Georgia)	Capital city (Tbilisi)
Type of vehicle and collection frequency	Waste collection trucks (up to 20 m³). Collection of waste in cities covers 100% of urban areas, but in mountainous village it drops to 18%.	–
% mixed waste or % source separated	Source separation has not been introduced in Georgia	
Processing		
Material recovery facility and recycling (%)	Formal recycling rate is 1%	–
Composting	Only a few local initiatives and pilot projects were fulfilled. One project in the west part of Georgia foresees 4000 T/y green waste composting from 2021.	N/A
Disposal – Open dump	Only 3 out of 60 official landfills are sanitary landfills (Tbilisi, Rustavi, and Borjomi)	
Disposal – Engineered/ Sanitary landfill	After disposal and compaction of mixed MSW on landfills, it is covered with soil material on a daily basis. Control of environmental parameters is conducted either according to the EIA (environmental impact assessment document) or Conditional Plans approved by the ministry of Environmental Protection and Agriculture of Georgia.	
Landfill gas – Collection	LG collection occurs in two active landfills of Rustavi and Tbilisi and in one closed landfill in Borjomi.	
Landfill gas – Flaring or LFG	First flaring system was installed on Rustavi Landfill in September 2020.	
Waste to energy	N/A	–
Major problems	• Old infrastructure and old system of WM in Georgia with minor changes • Lack of hazardous waste management system • Formation process of common database on WM of the country and local levels • Lack of EPR and PRO (Producer Responsibility Organization) systems • Very low private sector and/or informal sector participation in WM initiatives	

Table 3.7 (Continued)

Population	Country (Georgia)	Capital city (Tbilisi)
	• Lack of governmental programs to support recycling activities	
	• Legacy wastes	
	• Obsolete tariff system	
	• Lack of source separation	
Future needs	• Reform to the current WM system	
	• Proper collection and disposal of hazardous waste	
	• Common and reliable database on WM of the country, including local level data as well	
	• Social initiatives to involve producers, consumers, private sectors, and informal sectors in WM	
	• Considerable effort by government organizations to support and fund recycling activities, along with source separation	
	• Public awareness of recycling and source separation	

3.6 INDIA

Visva Bharati Barua is a Biotechnologist and an Environmental Engineer who completed her PhD from the Indian Institute of Technology Guwahati (IITG). She is currently a postdoc at the University of North Carolina at Charlotte (UNCC). Her research work focuses on combining biotechnological skills with environmental engineering. She basically specializes in converting biowaste into bioenergy and for her postdoc she is working on wastewater surveillance for COVID-19. She has nine publications in well-respected international journals and contributed a chapter to a book. She is the reviewer of 15 international journals of high repute and has received recognition for her outstanding work as a reviewer from Journal of Cleaner Production (Elsevier), November 2018 and Bioresource Technology (Elsevier), October 2017. She is the recipient of the Ministry of Human Resource Development (MHRD) fellowship her PhD research work and is the only Indian to receive an ISWA-SWIS full scholarship (2016). She has also received

best paper and poster presentation awards at national and international conferences. She loves conducting experiments and writing and believes in serving society with her bioenvironmental research contributions.

3.6.1 Introduction

Guwahati city is the heart of Northeast India. It has a population of 962 334 according to the census records of 2011 and generates 350-500 tons of municipal solid waste each day, of which 53.69% is biodegradable and 23.28% is recyclable (PCBA 2007). The amount of MSW generated in Guwahati has been steadily escalating and is attributed to the rapid growth of the population and the changes they have made in their lifestyles. The type and amount of MSW varies in line with the various waste generation sources in the city. The average waste generation in Guwahati city is 2.66 kg (5.85 lb) /day/household. Educational institutions, commercial establishments, hotels, marketplaces, street sweeping, and drain cleaning generate 129.59 tons (0.28 million lb) of waste per day. Although biomedical waste and industrial waste are not included in MSW, there are hospitals and a few industries that unofficially contribute to the total MSW. Construction and demolition waste generation is relatively less because most of the debris is utilized for filling low lying regions of the city. Industries like Indian Oil Corporation and the Guwahati refinery release only the treated waste, and the nonindustrial waste comes into the main solid waste stream of the city.

3.6.2 Collection

In Guwahati, MSW piles up on the streets due to the ignorance or lack of interest of common people. Individuals clean their own houses by disposing of wastes in their immediate surroundings, thus affecting the neighboring community and themselves. There is no segregation of wastes at the household level and all types of mixed wastes can be seen in the piles of MSW on the streets. This type of open dumping produces foul odors, breeds disease-carrying insects, and spoils the environmental aesthetics.

3.6.2.1 Collection and Processing of Wastes

The collection of solid wastes from Guwahati city is divided into two phases: primary and secondary collection.

1. **Primary Collection:** This stage consists of either direct disposal of the household waste by the owner into the dustbin provided by the Guwahati Municipal Corporation (GMC) or by door-to-door collection of household solid wastes. Nongovernmental organizations (NGOs) provide laborers with tricycle carts for door-to-door collection of solid wastes. They collect the wastes and unload them in secondary dustbins provided by the GMC. Such dustbin points

Figure 3.20 Truck carrying Guwahati city's solid waste to the Boragaon open dumpsite.

are available throughout the city at different points. The NGOs collect monthly charges from the households and commercial establishments for waste collection (Figure 3.20).

2. **Secondary Collection:** Secondary collection involves gathering the solid wastes that have been dumped into the dustbin points and conveying them to the open dumping yard. The institutional responsibility of the solid waste disposal system is assigned to the GMC. Garbage trucks appointed by the GMC are used to collect the wastes for the final disposal at the dumping yard. These garbage trucks make collections when the dustbins are full, which is approximately twice a week, during office hours, which creates a public nuisance. Some of the most commonly stated health issues are due to the uncollected wastes that are thrown onto the streets, thus clogging the drains, and leading to an increase in the number of mosquitoes.

3.6.3 Processing and Recycling

Scavengers redirect a substantial quantity of materials from the waste stream (Figure 3.21 a,b and c). They retrieve most of the recyclable paper, plastic, metal, and glass scrap from the Boragaon open dumpsite in Guwahati city and rudimentarily categorize and sell the collected recyclable commodities to retail scrap shops by weight. The sorting of waste is improved by the value or worth of the materials; however, the scavengers are paid very little for the commodities. A compost plant (Figure 3.21d and e) has been installed which has the capacity to produce 50 tons (0.1 million lb) of compost per day which is mostly supplied for use in the tea gardens of Assam, India.

Figure 3.21 (a) Scavengers rushing to pick up useful materials before they are covered (b) and (c) scavengers sorting out different types of waste (d) and (e) composting in Boragaon.

3.6.4 Final Disposal

The MSW collected from the city is finally openly dumped in the dumping ground of West Boragaon (Figure 3.22). The municipal trucks simply carry the wastes to the dumpsite and dispose of it without any processing, which has now become a health risk to the local people with the resultant pollution of air and water. During monsoons,

Figure 3.22 Boragaon, open dumpsite of Guwahati, Assam, India.

the condition of the open dump worsens and poses major challenges in upholding the approachability of the road and the wetland adjacent to it. The ever-increasing quantity of MSW disposed of indiscriminately in the Boragaon open dumpsite ultimately threatens the health and environment of the civil society.

3.6.5 Major Problems

The major problems of solid waste management in Guwahati are as follows:

- The inhabitants of Guwahati lack knowledge about the environmental and health hazards that can be caused by the improper disposal of solid wastes. Thus, waste reduction and segregation at the household level are not observed.

- Garbage trucks do not collect the waste at regular intervals. Moreover, they operate during busy traffic hours, causing a public nuisance.

- The open dumpsite of Guwahati (Boragaon) is situated adjacent to a large natural wetland and a Ramsar site, "Deepor Beel," which is an important destination for migratory birds. The leachate generated in the open dump pollutes the wetland (Figure 3.23), risking the life of innumerable aquatic organisms, migratory birds, and human beings (Barua 2016).

In order to fix the solid waste management scenario in Guwahati city, the following measures should be considered:

- Organic and inorganic waste should be properly segregated and separated at the household level.
- Efforts should be made to end the practice of open dumping and to upgrade to an engineered landfill with a gas-capturing facility.
- The scavengers should be provided with proper safety measures, technical expertise, and financial and legal advisory support.
- Garbage trucks should collect the wastes at regular intervals and early in the morning.
- The goal should be to keep the environment sustainable without compromising public health.

The summary of waste management in India is presented in Table 3.8

(a)   (b)

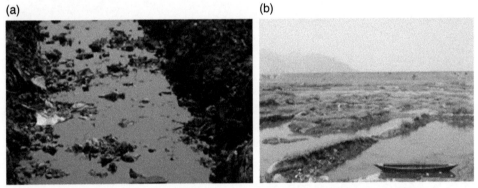

Figure 3.23 (a) Leachate generation from open dumpsite. (b) A large natural wetland adjacent to Boragaon open dumpsite.

Table 3.8 Waste Management in India.

Population	Country (India)	Capital (New Delhi)
Total population	1.339 billion	19.8 million
Urban population (%)	34	–
Rural population (%)	66	–
Waste generation		

Table 3.8 (Continued)

Population	Country (India)	Capital (New Delhi)
Waste generation/Year	62 million tons (136 billion lb)	3.2 million tons (7 billion lb)
Waste generation/ Person/Year	1.45 tons (3196 lb)	0.54 ton (1190 lb)
% organic waste	47–51	50
Collection – % Collected		
Type of vehicle and collection frequency	Tricycle cart for door-to-door collection and garbage trucks for final disposal. Frequency: once or twice a week	
% mixed waste or % source separated	Always mixed waste	Always mixed waste
Processing		
Material Recovery Facility and Recycling (%)	Scavengers	Scavengers
Composting	Yes	Yes
Disposal – Open dump	Yes	Yes
Disposal – Engineered/ Sanitary landfill	–	–
Landfill gas – Collection	–	–
Landfill gas – Flaring or LFG	–	–
Waste to energy	15 plants out of which 7 have been shut down and the rest are not functioning properly as mixed waste is utilized	3 plants
Major problems	• Lack of awareness	
	• Absence of source segregation and waste reduction	
	• Irregular waste collection	
	• Environmental issues	
Future needs	• Engineered landfill with MRF and WTE plants	
	• Support for scavengers	
	• Proper segregation at household level	

3.7 LEBANON

Nour Kanso works as an environmental consultant and a project coordinator for a solid waste facility. She graduated with a master's in environmental technology from Imperial College London. She is originally from Lebanon where she did her bachelor's in Environmental Health. Nour is very interested in exploring the different technologies related to the SWM sector.

3.7.1 Introduction

Lebanon, a country located on the east side of the Mediterranean Sea, is home to approximately 6.8 million people who mostly reside along the coast, particularly in the greater Beirut (capital) and Mount Lebanon regions (World Bank 2019; Worldometer 2020). The country faced a waste disposal crisis due to the closure of its biggest and most heavily relied-on landfill, the Naameh landfill. Lebanon generates about 2 700 000 tons (5.95 billion lb)/year of waste, which is projected to significantly grow given the high influx of Syrian migrants to the country. The region heavily relies on landfilling (52%), open dumping (32%), and burning, which result in severe environmental and health problems.

Lebanon's municipal solid waste mostly consists of organic material, estimated at 50%; recyclable materials estimated at 40%; and others, known as rejects, at 10% (Figure 3.24). The generation of MSW fluctuates between one area and another. In rural areas, the generation rate is at 0.8 kg (1.76 lb)/person/day, while in urban areas, where the generation is much higher, the rate ranges between 0.95 and 1.2 kg (2.09–2.65 lb)/person/day. Table 3.9 shows the generation rate

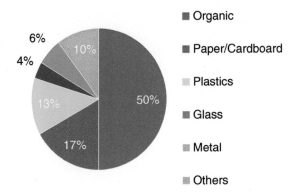

Figure 3.24 MSW composition in Lebanon (SWEEP-Net 2014).

- Organic
- Paper/Cardboard
- Plastics
- Glass
- Metal
- Others

Table 3.9 Waste generation per each region in Lebanon.

Area	Daily tonnage	Percentage of Country
Beirut (Capitol)	1550	21
Mount Lebanon	2590	35
South Lebanon	1180	16
North Lebanon	1180	16
Bekaa valley	890	12

Note: 1 metric ton = 2204 lb.

Source: Data from Lebanon Ministry of Environment (2017).

within each area. Mount Lebanon, which covers most of Lebanon's area, has the highest waste generation, accounting for 35% among other regions (PONDUS Consulting 2009; Ministry of Environment 2017).

3.7.2 Collection

MSW collection, prior to the crisis, was regarded as high, reaching a rate of 99% of collection in urban and rural areas. Curbside waste collection was used in most regions, where trucks collected an average of 2300 tons (5 million lb) per day and had a recovery rate of 6 to 7% per day (World Bank 2011).

3.7.3 Processing and Recycling

Different waste management practices (Figure 3.25) are presented here:

1. **Sorting and Bailing:** 44% of the generated solid waste in Lebanon is landfilled, 31% is dumped, and little is recycled or composted. Sorting of waste is mostly done manually; however, some regions have machines that do most of the work. Sorting is usually done by initially removing the big bulky solid materials, then sending the waste through a trommel screen to separate the organic fraction, which is collected into a separate bin and treated, where appropriate. The waste that remains behind is manually sorted to collect the recyclables that can be sources of income by baling and selling them to recycling companies/ centers (Ministry of Environment 2010; World Bank 2010).

2. **Recycling:** Based on data collected from the Ammroussieh and Karantine sorting facilities (Table 3.10), it was found that 33% of the waste recovered in Ammroussieh and 45% of the waste recovered in Karantine is cardboard. This is followed by plastic, which accounts for 18 and 17%, respectively, at the Amroussieh and Karantine sorting plants. Aluminum and tires are the items least recovered from

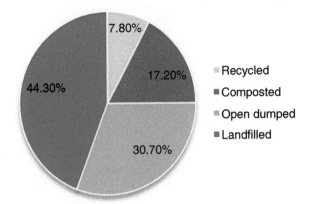

Figure 3.25 Different waste treatment methods in Lebanon (ISWA Lebanon 2019).

- Recycled
- Composted
- Open dumped
- Landfilled

Table 3.10 Average composition of recovered recyclable materials (CDR 2011).

Material	Composition by weight (%)	
	Amroussieh	Karantina
Plastic	18.2	17.4
Cardboard	32.9	45.1
Tin	28.5	20.7
Aluminum	0.8	1.0
Glass	13.8	11.0
Wood	4.5	4.0
Tires	1.3	0.8

the waste, given that the collected waste is from households rather than from industries. The two facilities sort about 41 000 tons (90.3 million lb) of recyclable materials annually and sell around 36 700 tons (80.9 million lb) of materials per year, which is about 90% of the total collected recyclable materials. Even though recycling is still sold at a low rate, there are markets that buy them.

3. **Composting:** As shown in Table 3.11, seven composting plants are operating across the Lebanese regions. The coral plant was selected for a case study and is briefly discussed because it accommodates most of the organic material generated from households. The coral plant treats about 110 000 tons (242 million lb) of organic material and generates about 41 000 tons (90.3 million lb) of compost. In general, the percentage of compost used by communities, industries, and individuals is around 22%, which can be increased if the composting process is properly supported and implemented.

Table 3.11 Amount of waste received in each composting plant.

Compost plant	Amount of waste received (tons/d)
Coral	300
Khaim	10
Ansar	10
Kherbetselem	15
Qabrikha	15
Ain Ebel	20
Bint Jbeil	50

Note: 1 metric ton = 2204 lb.

Source: Data from UNEP (2010c).

3.7.4 Final Disposal

There are few operating landfills across the country that can be described as sanitary; the others are more like controlled or uncontrolled dumping grounds. It was found that 410 000 tons (903 million lb) of MSW, including household, medical, and industrial waste, are disposed of in rivers, under bridges, and other places that are identified as open dumping areas (CDR 2009). According to the ministry, around 670 dumpsites were identified and are operating across the region, of which 27 are highly dangerous to the environment and in need of a lot of attention and action. The cost of environmental degradation due to waste dumping and burning is projected to be around 10 million dollars per year. The most well-known dumpsites in Lebanon are the "Tripoli" and "Saida" dumpsites (Figure 3.26) (Ministry of Environment 2010, 2013).

Tripoli's disposal site is located on the seashore of the city in the northeast region of Lebanon. The site receives about 350-400 tons (0.77–0.88 lb) of waste per day. The dumpsite occupies an area of 63 000 m² (6.3 hectares) and collects about 1000 m³ (35 314 ft³) of methane gas. Several enhancements have been made by the municipalities to improve the dump by collecting leachate, using it to hasten the decomposition of waste, and applying the daily cover to reduce the odors and deter rodents (Maasri 2012). The Saida dumpsite accommodates around 150 tons (0.33 lb) of solid waste per day from 15 municipalities. It was initially built to accommodate rubble and demolition waste; however, after a period of time, the dump began receiving all types of waste except medical.

The *Naameh Landfill*, now closed, was a sanitary landfill that accommodated about 2000 tons (4.4 million lb) of waste per day. Challenges of this landfill were mainly attributed to the high amount of waste disposal and the low capacity, improper operation, foul smell, and public protests (CDR 2011; Ministry of Environment 2011).

Figure 3.26 Saida's waste dump.

This eventually led to its closure and consequently created uncontrolled dumps and open burning of waste which led the country to its trash crisis. The *Bsalim Landfill* receives inert materials since the site is situated on an old quarry site in the western region of Lebanon. The Lebanese Council for Development and Reconstruction chose the quarry site for disposal of inert and bulky items as part of the restitution of the quarry and as a part of a comprehensive management plan for Lebanon. The landfill accommodates around 730 000 tons (1.6 billion lb) of waste with a total volume capacity of 1 000 000 m^2 (10.7 million ft^2) (CDR 2010). The *Zahle Landfill* is situated in the Bekaa valley area at Northwest region to Southwest region of Lebanon. The landfill was constructed to receive 150 tons (330 000 lb) daily and serve 15 of the 29 municipalities. The landfill accommodates around 43 000 tons (94.8 million lb) of waste per year. The cost of waste treatment and disposal at the Zahle SWM facility is approximately $40/t (UNEP 2007; CAS 2008).

3.7.5 Major Problems

The solid waste management system in Lebanon has several problems that involve irregular collection service, open dumping and burning, and the breeding of flies and vermin, all adding up to the improper control and handling of waste. Such public health, environmental, and management problems are attributed to poor execution of legislation, absence of cost recovery mechanisms and sustainable funding resources, lack of public involvement and awareness, and improper treatment and disposal.

The following measures can be undertaken to fix the problems:

- Improve the sorting and collection procedures.
- Strengthen the pertinent legislation.
- Implement a transparent and effective national waste management strategy in collaboration with stakeholders, NGOs, municipalities, government officials, and experts.

The summary of waste management practices in Lebanon is presented in Table 3.12.

Table 3.12 Waste Management in Lebanon.

Population	Country (Lebanon)	Capital city (Beirut)
Total population	6855713	3167839
Urban population (%)	78%	–
Rural population (%)	21%	–
Waste generation		
Waste generation/Year	2.7 million tons (5.95 billion lb)	~ Daily: 1550 MSW
Waste generation/ Person/Year	~ 0.95–1.2 kg (2.09–2.64 lb)/ capita/d in Urban areas ~ 0.80 kg (1.76 lb)/capita/d in Rural areas	~ 0.88–0.94 kg (1.94–2.07 lb) / capita/d
% organic waste	50% ~ 1320000 MSW	
Collection – % Collected		
Type of vehicle and collection frequency	Mostly daily to every other day in rural areas	Daily
% mixed waste or % source separated	Mostly mixed (no documented %)	Mostly mixed (no documented %)
Processing		
Material recovery facility and recycling (%)	7.8%	
Composting	17.2%	
Disposal – Open dump	30.7%	
Disposal – Engineered/ Sanitary landfill	44.3%	
Landfill gas – Collection	N/A	N/A
Landfill gas – Flaring or LFG	N/A	N/A
Waste-to-energy	N/A	N/A

(Continued)

Table 3.12 (Continued)

Population	Country (Lebanon)	Capital city (Beirut)
Major problems	• Poor Governance	
	• Poor Public Acceptance	
	• Upgrade and Improvement in Infrastructure	
	• Limitation in Human Resources Absence of cost recovery mechanisms and of sustainable funding resources	
Future needs	• Improved sorting and collection	
	• Strengthening of the legislation Implementing a transparent and accurate national waste management strategy with stakeholders, NGOs, municipalities, government officials, and experts	

3.8 MEXICO

Arely Areanely Cruz Salas received a bachelor's degree in environmental engineering and a master's degree in science and environmental engineering at the Universidad Autónoma Metropolitana in Mexico City. She works on research projects related to the management of urban solid waste, marine pollution (microplastics) in several matrices, waste pollution in protected natural areas, pollution caused by cigarette butts, compost, and vermicompost. She has carried out works related to the presence of microplastics on the beaches of the Gulf of Mexico and the Pacific. In addition, she has been the author and coauthor of articles in national and international journals. She is also a founding member of the project "Microplásticos en Ambientes Marinos," which has been carried since 2018 on Mexican beaches.

3.8.1 Introduction

Mexico has 32 federative entities and 130 million inhabitants (World Bank 2017). Mexico city (CDMX, by its initials in Spanish) is the capital and the most populous city of the Mexican Republic with a population close to 9 million (INEGI n.d., 2015). The total generation of solid waste (SW) in Mexico is over 42 million tons per year, while the per capita generation is equivalent to 1 kg (2.2 lb) per day

(SEMARNAT 2016). It was estimated that inhabitants of the capital city generated 12 998 tons of solid waste daily in 2017 with per capita generation of 1.37 kg (3.02 lb)/d (SEDEMA 2018).

In each region, the composition of the SW varies with socioeconomic status, culture, goods, and services (Cruz-Salas et al. 2018). In Mexico, the SW is composed of organics (52.4%), paper and cardboard (13.8%), plastics (10.9%), glass (5.9%), aluminum (1.7%), textiles (1.4%), ferrous metals (1.1%) and nonferrous metals (0.6%), and others (12.1%) (SEMARNAT 2015) as presented in Figure 3.27. Particularly in CDMX (Mexico City), the four most abundant categories are organics (34.93%), plastics (16.41%), sanitary (16.05%), and paper and cardboard (10.36%) (Cruz-Salas et al. 2018).

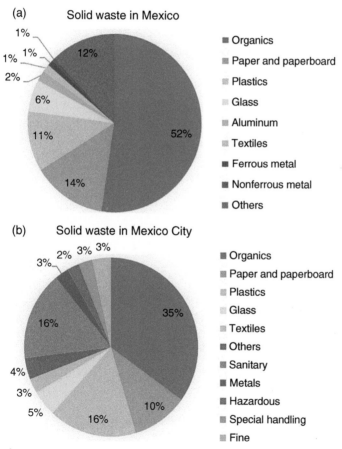

Figure 3.27 (a) Solid waste composition in Mexico. (b) Solid waste composition in Mexico city.

3.8.2 Collection

The efficiency of waste collection in Mexico has been reported to be approximately 93.4% of the total generated SW; however, this percentage is different in each of the states of the country. CDMX ranks fifth with 97% efficiency. At the national level, only 11% of the collection is selective (SEMARNAT 2015), while it is 46% in CDMX (SEDEMA 2018) in accordance with what is stipulated in the norm NADF-024-AMBT-2013 (Gaceta Oficial del Distrito Federal 2015).

The frequency of collection depends on the infrastructure and budget for the management of SW of each city. Generally, the metropolitan areas have a stronger infrastructure and a more generous budget, which means that the frequency of collection is higher in those areas than in the rural areas (SEMARNAT 2015). In the capital, waste is collected daily; however, the SW category is different every day of the week (Gaceta Oficial del Distrito Federal 2015).

In Mexico, there are 14 300 units of vehicles (Figure 3.28) that are used for collection, of which 61.73% are compactors, 34.02% are open boxes, and 4.25% are another type (INEGI 2014). In CDMX, 2566 vehicles are used for collection: 46.14%

Figure 3.28 Waste collection vehicles and workers.

are rear load, 21.67% are double compartments, 10.56% are dump trucks, 6.08% are rectangular, 1.05% are tubular, 0.82% are front load, and 13.68% are "others" (SEDEMA 2018).

3.8.3 Processing and Recycling

Around 9.6% of total generated SW (in volume) was recycled in Mexico in 2012 (Figure 3.29) and included paper and cardboard (32%), polyethylene terephthalate (PET, 15.8%), glass (13.8%), plastics (9.2%), metals (7.6%), and electronics and household appliances (5.1%) (SEMARNAT 2015). Only 4.55% (or about 1 million tons) of organic waste generated in Mexico undergoes treatment, and the type of technology used is unknown (INEGI 2012).

Two plants in CDMX receive 3858 tons of SW daily, of which 4% of the materials (mainly cardboard, paper, PET, and glass, among others) are recovered (SEDEMA 2018) – Figure 3.30. It has been estimated that more than 12% of the material recycled in CDMX is the product of informal activities (ONU Medio Ambiente 2018). Every day, 1374 tons (3.03 million lb) of organic waste are transferred from the Transfer Stations (TS) to 8 composting plants in Mexico City. The processing capacity of these plants is 923 996 tons (2.04 billion lb); however, only 99 803 tons of compost are produced from the 511 068 tons (1.12 billion lb) of SW that are received, and 6356 tons (14.0 million lb) of it are delivered to the delegations to improve the green areas (SEDEMA 2018).

Although CDMX has eight composting plants, their operation is not adequate, as there is a lack of water to moisten the piles, lack of infrastructure to flip the piles, lack of personnel, and too much waste to treat (Sánchez-Velasco et al. 2016;

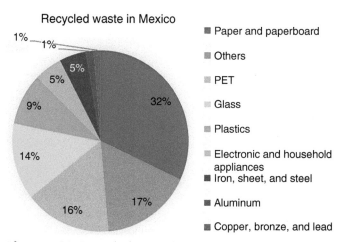

Figure 3.29 Recycled waste in Mexico in 2012.

Figure 3.30 PET recycling facility in Mexico.

Alvarez-Zeferino et al. 2017). In addition, there is a lack of monitoring, lack of control of parameters, and inadequate separation of the organics, all of which lead to the quality of the compost being compromised (Sánchez-Velasco et al. 2016).

3.8.4 Final Disposal

Sanitary landfills are one of the commonly used approaches for final disposal of SW. In 2013, 74% of the total SW generated was disposed of in sanitary landfills and controlled sites, while the remaining 26% was dumped in uncontrolled sites (open dumps) Figure 3.31 and Figure 3.32. With the exception of CDMX, almost all of the country uses sanitary landfills to dispose of the SW (SEMARNAT 2015). The SW that is generated in CDMX is sent to five sanitary landfills (El Milagro [48%], La Cañada [37%], Cuautitlán [8%], Chicoloapan [6%], and Cuautla [1%]) from neighboring states (Mexico and Morelos) (SEMARNAT 2015). Currently, 7862 tons (17.3 million lb) of SW per day are disposed of at these sites.

Figure 3.31 Open dumpsite in Chinampa de Gtza, Veracruz.

Figure 3.32 Open dumpsite in Naranjos, Veracruz.

3.8.5 Major Problems

The various irregularities that plague the waste management process in Mexico are listed later.

- The separate collection of recyclables (RS) highlights the low participation of the states (only 11% nationally) (SEMARNAT 2015). This may be because the RS collector truck service does not extend to distant or extremely poor localities, or if it does, it is infrequent.

- Another serious problem is the lack of personnel in the municipalities who have been trained to manage the SW. Little or no materials and equipment are available for use in the collection and treatment of SW and there are no economic resources for their acquisition (ONU Medio Ambiente 2018).

- Even though all of the entities in Mexico have disposal sites (SEMARNAT 2015), they often fail to meet the specifications set by the NOM-083-SEMARNAT-2003 norm for the establishment of sanitary landfills (Gobierno del DF 2003). Therefore, the disposition of the SW is done in ways that harm the environment, workers, and inhabitants surrounding the sites. The CDMX does not have sanitary landfills; therefore, the SW is disposed of in nearby sites, which incurs an increase in the cost of transferring it from transfer stations to landfills (SEMARNAT 2015).

- In CDMX, there is little societal participation in separating the RS from the source, and those who do separate it encounter problems in delivering it to the collection truck on the day specified.

- Most compost plants that treat organics have operational deficiencies that are associated with the lack of resources and result in the product (i) failing to reach maturity, (ii) lacking optimum parameter values, (iii) being poor quality, and (iv) unable to provide environmental and economic benefits (Alvarez-Zeferino et al. 2017). In addition, the accumulation of wooden logs that could cause fires pose potential risks to the environment and workers (Sánchez-Velasco et al. 2016).

- Few recyclable materials reach the selection plants in CDMX because informal garbage collectors gather them from the household collection service. This diminishes the economic benefits in such infrastructures. Of all the materials that can be recovered in the CDMX, PET is the most recycled, but it only represents 2% of its total plastic waste (Cruz-Salas et al. 2018). The closest plant for recycling PET is PETSTAR, in Atlacomulco, State of Mexico (PETSTAR 2019).

The following measures should be considered for addressing the problems associated with solid waste management in Mexico.

- The management of RS in Mexico shows various inefficiencies. When formulating the Mexican legislation, it would be advisable to establish who is responsible for verifying compliance and enforcing fines on those who are in violation.
- Municipal personnel who are responsible for the management of SW should receive training that will enable them to make better decisions regarding public cleanup, waste treatment, and final disposal.
- A fee that is based on the income level of each zone should be levied to provide resources that are needed for effective waste management.
- The citizens should be made environmentally aware of the importance of their waste being managed in a safe and sustainable manner. They should be encouraged to separate the valuable waste from the source at the household level. For the organic waste, ideally each municipality could establish a small compost or vermicompost plant for its treatment and provide workshops to teach the citizens how to make homemade compost.
- For other recyclable materials, the municipalities could offer each citizen a voucher equivalent to the amount of SW they deliver to the collector truck, and the vouchers could be accumulated and used to pay property taxes.

The summary of waste management system in Mexico is presented in Table 3.13

Table 3.13 Waste Management in Mexico.

Population	Country (Mexico)	Capital (Mexico City)
Total population	130 000 000	8 985 339
Urban population (%)	80.2	99.5
Rural population (%)	19.8	0.5
Waste generation		
Waste generation/Year	42 000 000 tons (92.6 billion lb)	4 744 270 tons (10.4 billion lb)
Waste generation/ Person/Year	365 kg (805 lb)	500.05 kg (1102.4 lb)
% organic waste	52.4%	34.93%
Collection – % Collected	93.4%	97%

(Continued)

Table 3.13 (Continued)

Population	Country (Mexico)	Capital (Mexico City)
Type of vehicle and collection frequency	Vehicles: compactor (61.73%), open box (34.02%), and others (4.25%) Frequency: variable	Vehicles: rear load (46.14%), double compartment (21.67%), others (13.68%), dump truck (10.56%), rectangular (6.08%), tubular (1.05%), and front load (0.82%) Frequency: Daily
% mixed waste or % source separated	Mixed waste: 89% Source separated: 11%	Mixed waste: 54% Source separated: 46%
Processing		
Material recovery facility and recycling (%)	Facilities: there are some but the exact number is not available Recycling: 9.6% of total waste generated	Facilities: there are two selection plants Recycling: 4% of waste received
Composting	Information not available	Only 30.8% of total organic waste received in the composting plants is used
Disposal – Open dump	26% of total waste generated	Not available
Disposal – Engineered/ Sanitary landfill	74% of total waste generated	60% of total waste generated
Landfill gas – Collection	7800 m³/h	Not applicable
Landfill gas – Flaring or LFG	Installed capacity: 60.727 MW Generation of electricity: 61.52 GWh/yr	Not applicable
Waste to energy	A Waste to Energy facility that will treat 1.6 million tons of household waste per year (3.80% of the total waste generated) is under construction in Mexico City.	
Major problems	• Untrained personnel • Lack of material and equipment for collection and treatment • Lack of economic resources and noncompliance with legislation for sanitary landfills	• Noncompliance with the waste norm NADF-024-AMBT-2013 • Less organic composting and operational deficiencies in the plants • Significantly less recovery of recyclables and recycling facilities

Table 3.13 (Continued)

Population	Country (Mexico)	Capital (Mexico City)
Future needs	• Perform a current diagnosis of the generation of RS at the national level • Supervision of the compliance of the Mexican legislation in the area of RS • Training of the municipality staff • Application of a specific fee for the RS collection service • Installation of municipal composting plants • Purchase of recyclables from citizens	• To be rigorous before the fulfillment of the norm NADF-024-AMBT-2013 for separation of RS and the norm of production of compost NADF-020-AMBT-2011

3.9 PAKISTAN

Maria Ajmal is passionate about exploring strategies for implementing the concept of integrated waste management in developing countries. She is an advocate of source segregation of waste, and her research has led to her experience in quantification and characterization of municipal solid waste, simultaneously understanding the importance of key determining factors such as the roles of the community and other waste producers in an effective management plan. She is further excited to develop an insight into enumerating the contributions of sustainable waste management to climate change mitigation, utilizing the dynamics of environmental policy and economics.

3.9.1 Introduction

Pakistan is the fifth most populous country in the world and is home to 207 million people (PBS 2017). This population produces about 20 million tons of waste per year, an amount that is growing at an annual rate of 2.4% (Korai et al. 2017). According to the

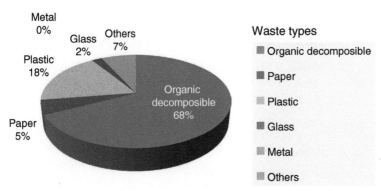

Figure 3.33 Typical waste composition in Pakistan.

World Bank Report, "What a Waste? A Global Review of Solid Waste Management," this rapid increase in waste generation will result in doubling the current waste generation by 2025. The predicted rise is mainly attributed to the mounting population, combined with urban sprawl and economic growth (6% annual GDP increase) (PakEPA 2005). Fulfilling the generalized profile of almost all developing countries, Pakistan generates a relatively higher percentage of organic decomposable waste than other types of waste (Figure 3.33). Plastic is the second most prevalent type of waste.

Hazardous waste is another type of waste being produced in Pakistan from sources like hospitals, the agriculture sector, industries, commerce, and households. Unfortunately, due to lack of data and improper disposal procedures, it is treated as ordinary waste, with most of it being dangerously dumped in the tributaries of the Indus River and/or the Arabian Sea. The guidelines for the management of such waste by the government of Pakistan are purposeless due to their lack of implementation.

3.9.2 Collection

Collection, despite being the most integral part of waste management, is poorly managed in the majority of the cities in Pakistan. In urban and semi-urban areas, where population is lesser than rural, the average waste collection percentage is better, ranging between 50 and 69% (Mahar et al. 2007). One of the reasons for not reaching optimal collection is incorrectly positioned formal collection bins that are also insufficient in number as compared to the population and its generated waste volumes. Another reason is that the size of the collection vehicles is not appropriate for the poorly designed cities. In congested areas, the vehicles provided by the municipal corporation are banned from the hearts of the residential areas. As a result, the waste accumulates and lies undisturbed in areas in the communities for prolonged periods of time. In very poor communities, there is no removal mechanism, and the waste is just disposed of by informal scavenging, natural degradation, dispersion, and illegal burning.

Figure 3.34 Primary waste collection vehicles.

Figure 3.35 Secondary waste collection vehicles.

About 5% of the waste is separated by households or commercial producers before getting mixed or picked and separated from mixed waste dumps by waste scavengers.

Waste that does get collected is managed in two phases (Figure 3.34 and Figure 3.35). The first phase employs the use of handcarts, donkey pull carts, and auto rickshaws that are usually owned by independent operators and workers (Figure 3.36) collect waste more frequently either daily or on alternate days. The secondary phase entails using government-owned open trucks, tractors/trolleys, and arm-controlled trucks (PakEPA 2005; Pakistan-Waste Management 2018). This occurs less frequently, i.e. once or twice a week.

3.9.3 Processing and Recycling

Heavily populated urbanized cities are usually more developed and have material recovery and composting facilities because of the significant quantities of waste that they generate. For instance, Lahore, the provincial metropolitan area of Punjab, has a large waste recycling and processing industry. The Lahore Waste Management

Figure 3.36 Waste collection workers.

Company (LWMC) works with the support of the Punjab Information Technology Board to handle the collection, transportation, and disposal of 5000 tons (11 million lb) of waste produced daily. The revolutionary achievement of LWMC, however, is the establishment of a sanitary landfill that has been designed for material recovery. The company is responsible for providing 500–800 tons (1.1–1.8 million lb)/d of waste for a composting company (Figure 3.37) called Lahore Compost Limited (LCL) and about a 1000 tons daily for a refuse-derived fuel plant owned by DG Khan Cement (Rizvi n.d.). Another industry called Green Earth Recycling (Figure 3.38 and Figure 3.39) processes approximately 500 tons (1.1 million lb) of recyclables daily from various cities in Punjab

Figure 3.37 Composting facility.

(a)

(b)

Figure 3.38 Plastic and paper recycling (a) Middlemen and (b) facility.

Figure 3.39 Iron and steel recycling facility.

to make high-quality building materials and furniture (Green Earth Recycling n.d.). The facility also works in collaboration with multinational companies such as Tetra Pak in order to help them achieve a corporate socially responsible image and compensate them for being waste producers.

Other such initiatives that are being planned by Rawalpindi Waste Management Company (RWMC) will involve source separation of waste and community awareness that will lead to waste-to-energy projects (WMW 2015). In other cities of the country, informal recycling by scavengers is a huge business. This sector is mainly composed of religious minorities (Hindus, Christians) and Afghan refugees or migrants who reduce the amount of waste dumped on the streets while generating a little income for themselves (US $ 0.6 to 1.1/d) (PakEPA 2005). Unfortunately, this sector is unable to expand or perform optimally as the setup does not fulfill the state institutions' criteria (Ahmed 2018).

3.9.4 Final Disposal

Relatively improved dumping sites, sometimes claimed as sanitary or engineered landfills, are another of the "privileges" enjoyed by those living in the developed areas of Pakistan (Figure 3.40 and Figure 3.41). Unfortunately, however, most of the metropolitan cities are served by the Mehmood Booti Landfill of Lahore and the dumpsites located in Surjani Town's Jam Chakhro and the Deh Gondal Pass in Karachi, which have reached or are fast approaching their saturation point. In one city, the administration is looking at the prospect of waste-to-energy that could potentially generate 200 megawatts of electricity for energy-crisis-stricken Pakistan. Unfortunately, owing to the lack of research on the changing composition of waste, such initiatives rarely make it beyond the idea phase. Even partnerships with companies from China and Turkey have not been able to achieve scientific disposal of waste that results in resource recovery on a large scale. The situation in smaller cities is graver, as uncovered open dumping on rented/leased land space is prevalent and the imprudent waste disposal practices lead to numerous environmental problems.

Figure 3.40 Dumpsite in Karachi.

Figure 3.41 Dumpsite in Islamabad.

3.9.5 Major Problems

The most significant impacts created by the mismanagement of waste in Pakistan are caused by its improper disposal. Since open dumping is the most convenient solution available in most cities, it is widely practiced, and the carelessly positioned dumps give off odors and disturb the aesthetic pleasance of the area. More importantly, they become home to pathogens and disease-causing vectors, giving rise to diseases such as malaria and cholera. It has been estimated that about 5.2 million people, including children, are fatally impacted by waste-related diseases annually (Hussain and Mushtaq 2014). Most of the dumpsites are located near water bodies, on flood plains and in ponds; consequently, water contamination occurs due to the direct contact with waste or the seeping of leachate through the soil. More urbanized cities are also running out of space in which to dump waste, further aggravating the problem.

Waste management in Pakistan is victim to barriers that need to be reversed, including those listed later.

- Prioritizing the issue: The responsible authorities are either unaware of or in denial about the problems related to waste management and need to be educated about the importance and urgency of addressing them.

- Ensuring review and implementation of regulations: Despite the presence of a regulatory system and rules for the management of solid waste in the country (Policy and Regulations on Solid Waste Management 2010), the lack of implementation makes all such efforts null and void. Institutions need to be organized and strengthened so that they can ensure constant review and implementation of the rules.

- Public–private partnerships for capacity building: A collaboration of the public and private sectors can help increase the capacity and improve the efficiency of the system.

- Strengthening the minorities, managing waste: Due to the socially unacceptable nature of the service, waste management has been designated to only a few ethnic minorities who are living in poverty and strive to make a living. A system needs to be developed that will support these workers by providing them better employment prospects and upgradation of their social status with provision of medical facilities, PPEs, etc.

- Innovative research for sustainable solutions: Targeted scientific research is essential to finding comprehensive solutions to the problems of waste management. Since Pakistan is a low-income economy, resource recovery from waste must be considered as a more sustainable approach and a step toward achieving a circular economy model. Solutions leading to resource recovery have the potential to solve the problems of solid waste mismanagement and to contribute to economic development.

- Creating environmental awareness among the community: Community participation is one major factor that can be utilized in favor of the management. The community needs to be educated about the negative impacts of solid waste mismanagement so that they are willing to participate in activities such as waste minimization and source segregation that have the potential to dramatically improve the efficiency of resource recovery.

The summary of waste management in Pakistan is presented in Table 3.14.

Table 3.14 Waste Management in Pakistan.

Population	Country (Pakistan)	Capital (Islamabad)
Total population	207 774 520	2 001 579
Urban population (%)	36	50
Rural population (%)	64	50
Waste generation		
Waste generation/Year	20 042 000 tons (44.2 billion lb)	201 171–221 289 tons (443 mill lb–486 mill lb)-
Waste generation/ Person/Year	83.95–222.65 kg (185–490 lb)	103.295–223.745 kg (227.7–493.2 lb)
% organic waste	68%	64%
Collection – % Collected	50–69%	60–70%
Type of vehicle and collection frequency	Primary collection: Handcarts and donkey pull-carts Secondary collection: open trucks, tractor/trolley systems, and arm roll containers/trucks Frequency: primary – almost daily Secondary – once or twice a week *Note: The frequency may differ from area to area*	Hydraulic refuse packers (garbage compacting vehicles) & skip lifters, dump trucks tractor trolleys for green waste, and construction debris Frequency: Once a day from 08.00 A.M to 04.00 P.M
% mixed waste or % source separated	Mixed waste: 90–95% Source separated: 5–10% (includes products separated and sorted by scavenger)	
Processing		
Material recovery facility and recycling (%)	Mostly exists in metropolitan cities. Of the recyclable fraction (about 25% of total mixed waste), roughly 20% of it actually goes to the recycling facilities. Some of it is source separated by households informally and bought by waste buyers (kabadias), while the rest is retrieved by scavengers from dumps.	Small-to-medium scale recyclers exist that rely on sorting performed by private operators. About 20% of the recyclable fraction is processed.

(Continued)

Table 3.14 (Continued)

Population	Country (Pakistan)	Capital (Islamabad)
Composting	Only in limited number of metropolitan cities like Lahore	None
Disposal – Open dump	Widely practiced	Widely practiced
Disposal – Engineered/ Sanitary landfill	Mostly in metropolitan cities like Lahore and Karachi	Present
Landfill gas – Collection	None	None
Landfill gas – Flaring or LFG	None	None
Waste to energy	None	None
Major problems	• Improper disposal of waste • Spread of diseases leading to public health issues • Contamination of water bodies • Reduced land availability for dumping	
Future needs	• Prioritization of solid waste management by authorities • Ensuring review and implementation of regulations • Public–private partnerships for capacity building • Strengthening the community that is managing waste (minorities and refugees) • Innovative research for sustainable solutions • Creating environmental awareness among community	• Revision and proper implementation of SOPs • Efforts to increase the potential of recycling system • Incorporating public awareness through educational institutions

Note: Data were gathered from literature review and personal communication with A. Sheikh, Owner, Qureshi Plastic Factory on 15 May 2019.

3.10 PORTUGAL

Soraia Taipa has a background in environmental engineering and project management. As an innovation manager at a sustainability driven waste management company. LIPOR (Portugal), she aims to contribute to the development of sustainable businesses that see potential in nature-based solutions. LIPOR is a pioneer in sustainable waste management solutions and promotes win-win partnerships across the world. With 10 years of experience in the environmental sector, Soraia Taipa has enabled research and outreach projects in academia, private, and nonprofit sectors. She has worked on diverse sustainable goals by working on a circular economy, ecosystem services, local strategic plans, legal frameworks, reforestation, and environmental awareness. Soraia Taipa is an active member of several professional networks. She founded and chaired Smart Waste Portugal Young Professionals Group and was co-leader of Research & Innovation by International Solid Waste Association Young Professionals.

3.10.1 Introduction

Portugal is a European country with 10 million inhabitants that is located southwest of Europe (Instituto Nacional De Estatistica 2021). The urban population has been increasing for the last several decades and now almost 50% of the population lives in urban areas, close to the sea.

The production of MW has been increasing since 2014, along with an improvement of the economic situation after a severe economic crisis. In 2017, each Portuguese generated 487 kg of waste per year, the exact average of European Union Countries, corresponding to a daily capitation of 1.33 kg/ (inhab.day), 36% of which was organic waste.

According to Portuguese Environment Agency, Waste management in Portugal is ruled by the European Waste Framework Directive, which distinguishes the hierarchy of waste, the extended producer responsibility, and a life cycle perspective (*Agência Portuguesa do Ambiente* 2021). An Integrated System for the Management of Packaging (SIGRE) was developed so that packaging producers can transfer their responsibility to recover and treat packaging waste to a licensed organization licensed by PRO EUROPE (Packaging Recovery Organization Europe). Until 2016, Portugal had only one company licensed by PRO EUROPE, Sociedade Ponto Verde; currently, there are three such companies.

Portugal created Municipal Waste Management Systems in 2017, which is responsible for complying with regulations and EU targets for waste management and for creating the infrastructures for sorting and shipping waste to recycling centers or its final destination. 2020 European Union targets for waste management are deployed in Portugal through The Strategic Plan for Municipal Solid Waste, PERSU 2020 (https://apambiente.pt/_zdata/DESTAQUES/2014/PERSU2020_Relatorio_Ambiental_Final.pdf). Municipal Waste Management Systems developed initiatives to comply with PERSU 2020 targets that were to be reached by 31 December 2020, by achieving:

- a minimum reduction of waste production per inhabitant of 10% by weight of that produced in 2012
- a minimum overall weight increase of 50% in the preparation for reuse and recycling of MW, including paper, cardboard, plastic, glass, metal, wood, and biodegradable MW
- a 35% reduction in the amount of biodegradable MW deposited in landfills in 1995.

3.10.2 Collection

Municipal waste in Portugal is currently based on separate collection of packaging waste with three recycling bins: a blue bin for paper and cards, a yellow bin for plastics and metals, and a green bin for glass. Collectively, the bins are referred to as an "Ecopoint." The Decree-Law 366-A/97 of December 20th on waste gives reference values on density of ecopoints (200 inhabitants per ecopoint) and distance between ecopoints (maximum distance of 200 m between ecopoints). The number of ecopoints has been increasing significantly and reached 43 600 in 2017.

Other recoverable waste streams are door-to-door and Ecocenters. Separate collection of organic waste is made on door-to-door streams. An increase of the quality of all recyclables, including organic waste, has been observed since the onset of door-to-door separate collection.

Street collection of waste, separated or mixed, is mainly done by big trucks (Figure 3.4a). Small vehicles are used for door-to-door streams. Mixed waste is collected almost daily (Figure 3.42c), and separated waste collection is collected one or two times a week (Figure 3.42b); in urban areas, it may be collected three or four times a week.

3.10.3 Processing and Recycling

According to Pordata (https://www.pordata.pt/DB/Municipios/Ambiente+de+Consulta/Tabela), approximately 16.5% of the waste is collected, sorted, and shipped to a recycler (Figure 3.43). The bulk collection is sent for mechanical treatment and/or mechanical

(a)

(b)   (c)

Figure 3.42 Waste collection (a) vehicle, (b) collection of recyclables, (c) mixed waste collection.

and biological treatment to recover the recyclables or perform another recovery process. A total of 38% of MW is prepared for reuse and recycling; however, only 11% of the total MW is effectively recovered (2017 data). Organic waste is managed by composting/anaerobic digestion and represents 9.8% of all of the MW.

3.10.4 Final Disposal

Bulk collection still represents the bigger fraction, with two main management destinations: waste incineration with energy recovery (21%) (Fugre 3.44) and engineered landfills (32%) (2017 data) (Figure 3.45).

3.10.5 Major Problems

A brief analysis of the evolution of waste management in Portugal for the last five years shows a decrease in the growth rate of separately collected MW and a continued large amount of direct landfill deposition.

Figure 3.43 Waste sorting center.

The ambitious targets set for the quality and quantity of separately collected recyclable waste can only be met if the people invest their time and efforts. PAYT (Pay as You Throw systems) have some pilot operations running and could be used as a way to motivate the general population to become involved.

Portugal established an Action Plan for Circular Economy 2017–2020 to foster an increase in the economic value of recyclables and by-products of MW and to increase the efficient use of their natural resources. To strengthen the transition with a closing-the-loop approach, the plan leverages a decrease in the importation of materials and waste prevention and an increase in upcycling the value of resources in circulation. The plan consists of the following:

- acquiring knowledge by researching innovative ways to create solutions.
- Product – consumption – knowledge approach for a flourishing, responsible, dynamic, and inclusive society (design, repair, reuse, circular markets, and education).

(a)

(b)

(c)

Figure 3.44 Waste-to-energy plant (a) mixed waste, (b) plant, (c) fly ash.

Figure 3.45 Bottom ash landfill.

- creating an inclusive and resilient economy by eating without waste, finding new ways to utilize waste, and regenerating resources.

The summary of waste management practices are presented in Table 3.15.

Table 3.15 Waste Management in Portugal (Lisboa a Compostar 2021; REA 2021).

Population	Country (Portugal)	Capital (Lisbon)
Total population – 2017	10 291 027	506 088
Urban population (%) – 2017	43.6 and 30.3% on middle term	100% urban
Rural population (%) – 2017	26.1%	0%
Waste generation		
Waste generation/ Year – 2017	5 007 000 ton (11 billion lb)	321 366 ton (0.71 billion lb)
Waste generation/ Person/Year – 2017	487 kg (Europe average) (1073 lb)	635 kg (1399 lb)
% organic waste – 2017	36.56%	40%
Collection – % Collected		
Type of vehicle and collection frequency	Mainly big trucks Frequency: bulk collection – daily Separate collection – 1–2 times a week	
% mixed waste or % source separated – 2017	Bulk collection: 83.5% Separate collection: 16.5%	Bulk collection: 73% Separate collection: 27%
Processing		
Material recovery facility and recycling (%) – 2017	Material recovery: 11% Reuse and recycling: 38%	–
Composting	Composting/anaerobic Digestion: 9.8%	Data for VALORSUL que company that manages waste of Great Lisbon, including the municipally of Lisbon Composting: 4%
Disposal – Open dump	None	None
Disposal – Engineered/ Sanitary landfill	2017 – Landfill: 32%	Data for VALORSUL que company that manages waste of Great Lisbon, including the municipally of Lisbon Sanitary Landfill: 22%
Landfill gas – Collection (2014)	Data found: energy pro- duction through Landfill Gas in Portugal – 240 GWh 96% of the total energy production through biogas	No data found

Table 3.15 (Continued)

Population	Country (Portugal)	Capital (Lisbon)
Landfill gas – Flaring or LFG	No data found	No data found
Waste to energy	Energy recovery: 21%	Data for VALORSUL que company that manages waste of Great Lisbon, including the municipally of Lisbon Energy recovery: 60%
Major problems	• Slow growth in the rate of separately collected MW	
	• High direct landfill deposition	
Future needs	• Circular economy approach	
	• Increase separate collection	
	• Increase door-to-door collection systems	
	• Implement PAYT – Pay as You Throw	
	• Divert organic waste from landfill	
	• Prioritize waste prevention	
	• Dissociate waste production from economic growth	

3.11 SERBIA

Dusan Milovanović is a solid waste expert with more than 15 years' experience with activities related to solid waste engineering activities in Serbia and Southeast Europe. Mr. Milovanović has coordinated nationally- and internationally focused waste-oriented projects supported by the Serbian Ministry of Education and Science, European Commission, NATO, US Environmental Protection Agency, and the Climate and Clean Air Coalition/ UN Environment. His work has included identification and assessment of dumpsites, remediation planning, application of GIS technology to disposal site planning, and economic analysis of remediation. He also has a strong focus on qualitative and quantitative modeling and analysis of leachate and landfill gas production at disposal sites in the Vojvodina region of Serbia. He has strong hands-on experience with materials recovery and recycling activities, including works estimating the potential for generation and processing of urban organic waste in both Serbia and Bosnia Herzegovina. Mr. Milovanović is a member of various professional organizations including ISWA and SeSWA and is a strong communicator. He has experience in a wide variety of Serbia and internationally focused waste activities.

3.11.1 Introduction

Serbia has a total population of 7 186 862 (excluding Kosovo) as per the 2011 census. The majority of the population (about 59.4%) resides in urban areas, with almost 18.8% of them living in the capital city of Belgrade. Belgrade is the only city in Serbia with more than one million inhabitants. The population of Belgrade including suburbs is 1 659 440 with 81.04% living in urban areas. According to the Official Report issued by the Serbian Environmental Protection Agency (SEPA 2017), "Waste Management in Republic of Serbia from 2011-2017," generation of municipal solid waste in 2017 in Serbia was 2 150 000 tons (4.7 billion lb), representing a generation rate of 0.3 ton (661 lb) per person per year, or 0.84 kg (1.85 lb) per person per day. In Belgrade, 660 208 tons (1.5 billion lb) of municipal solid waste are produced every year, with each person generating around 0.4 tons (881 lb) per year. Various studies have been conducted in Serbian municipalities to determine the composition of the waste and the approximate share of the different waste streams. Organic waste (food and green waste) has been found to represent a higher fraction in the waste stream, with an average share of about 49.54% in Serbia. In the city of Belgrade, organic waste constitutes about 45.08% of the total waste generated. Other waste types include paper and cardboard, plastics, glass, composite packaging, metals, tires, e-waste, etc.

3.11.2 Collection

No source separation system has been implemented yet in the municipalities of Serbia except for in some pilot projects in a few cities. Therefore, the majority of the collected waste is mixed. Based on the report by the Serbian Environmental Protection Agency (SEPA), only 0.3% of the MSW was separated in 2017. The percentage of waste collected in Serbia was reported by SEPA to be 83.7% for the period of 2011–2017, with approximately 1 800 000 tons of waste collected and disposed of in the landfill during that period. The MSW collection rate for the city of Belgrade is 95%, as per the information provided by Public Utility Company City Sanitation (JKP Gradska Cistoca).

The frequency of waste collection and types of vehicles used vary from municipality to municipality. The collection trucks make more frequent visits in urban areas than in rural areas (Figure 3.46). In Belgrade, public utility companies (PUC) use power-press garbage trucks to collect MSW from urban and commercial zones. Wastes are collected on a daily basis from commercial zones (industrial and businessparks, shopping centers, etc.) and collective housing areas and are collected weekly for individual households, which are generally in rural parts of the city. A similar collection system is practiced in many cities of Serbia.

Figure 3.46 Waste collection workers and vehicle.

3.11.3 Processing and Recycling

Most of the collected wastes are taken to the landfill, and small amounts are delivered to MRF facilities that are located in the vicinity of the landfills. MRF facilities are only available in six municipalities of Serbia (Novi Sad, Obrenovac, Uzice, Pancevo, Leskovac, and Sremska Mitrovica). PET, paper and cardboard, glass, and metal are some of the materials recovered from the waste in MRF facilities (Figure 3.47a, b). According to the EuroStat (2021) data, only 1% of MSW is recycled in Serbia, while in Belgrade it is 1.1% (Figure 3.47c)

There are no composting plants or waste treatment facilities in Serbia; however, several pilot projects are composting green waste. Plans for the construction of several MBT, AD, and composting plants are ongoing, and construction is expected to start soon in a few places.

The Landfill Directive, which was established in 1999, has been very important for processing biodegradable MW. It requires the member states to set up national strategies for reducing the amount of biodegradable MW going to landfills within a specific timeframe by means of recycling, composting, biogas production, or materials/energy recovery. As an EU candidate country, Serbia is obligated to transpose EU directives into the National legislation; thus, it is also required to adjust the management of biodegradable MW.

3.11.4 Final Disposal

Approximately 95% of the MSW is disposed of at the MSW landfills and open dumps. A study conducted by the Department of Environmental Engineering, Faculty of Technical Sciences, University of Novi Sad determined that there are around 3500 waste disposal

(a)

(b)

(c)

Figure 3.47 (a) and (b) Material recovery facility and (c) recycling facility.

Figure 3.48 Landfill working face.

locations in Serbia (unsanitary MSW landfills and open dumps) and around 133 in the city of Belgrade.

The National Strategy for Waste Management from 2010 to 2019 (Official Gazette of RS – 29/2010) foresaw the closure and remediation of all current unsanitary MSW landfills and the construction of 29 regional sanitary landfills, along with recycling yards and transfer stations. According to the Strategy and the Law on Waste Management, it was proposed that regional centers (that covers at least 250 000 inhabitants), based on the construction of sanitary landfills with additional treatment technologies, be established as the optimal solution for waste management in Serbia; however, only 10 sanitary landfills have been constructed since 2002 and in 2017, 460 488 tons (1.01 billion lb) of MSW (21.4%) were disposed of in sanitary landfills (Figure 3.48).

3.11.5 Major Problems

The major problems of solid waste management in the cities of Serbia are as follows:

- The lack of a source separation system.
- The lack of a systematic approach, regulations, and public awareness regarding the need to separate the waste at source which are crucial for sustainable waste management.
- The open dumps and unsanitary MSW landfills pose significant threats to human health and the environment.
- There is an immediate need to address the critical issues of open dumps and unsanitary landfills, such as the installation of an LFG collection system, with either flare or LFGE application, closure, or land remediation.

The following should be done in Serbia to deal with the waste management issues:

- Implement source separation in the household and commercial sectors.
- Raise public awareness of the urgent need for source separation and adequate waste disposal.
- Close or remediate existing unsanitary municipal landfills and open dumpsites, along with constructing sanitary landfills.
- Construct biodegradable waste treatment plants.
- Construct waste treatment facilities and recycling yards.
- Construct LFG collection systems and leachate collection systems and treatment.

The summary of waste management in Serbia is presented in Table 3.16

Table 3.16 Waste Management in Serbia.

Population	Country (Serbia)	Capital (Belgrade)
Total population	7 186 862	1 659 440
Urban population (%)	59.44	81.04
Rural population (%)	40.56	18.96
Waste generation		
Waste generation/Year	2 150 000 tons (4.7 billion lb)	660 208 tons (1.5 billion lb)
Waste generation/Person/Year	0.30 ton (661 lb)	0.40 ton (881 lb)
% organic waste	49.54%	45.08%
Collection – % Collected	83.7%	95%
Type of vehicle and collection frequency	Varies from city to city and in urban and rural areas	Power-press garbage trucks Frequency: Individual households – 1/wk Commercial zones and collective housing – everyday
% mixed waste or % source separated	Source separated:0.3%	No info
Processing		
Material recovery facility and recycling (%)	6 MRF facilities Recycling: 1%	No MRF Recycling: 1.1%
Composting	–	–
Disposal – Open dump	3500 nos.	133 nos.
Disposal – Engineered/ Sanitary landfill	10 nos.	0

Table 3.16 (Continued)

Population	Country (Serbia)	Capital (Belgrade)
Landfill gas – Collection	–	–
Landfill gas – Flaring or LFG	–	–
Waste-to-energy	One facility Municipal waste and other waste streams used in cement factory	–
Major problems	• No source separation system	
	• Lack of public awareness of need for adequate waste disposal	
	• Environmental issues with existing unsanitary municipal landfills and open dumpsites	
	• Lack of LFG collection and utilization systems	
Future needs	• Implementation of source separation in households and commercial sector	
	• Raising public awareness of the importance of source separation and adequate waste disposal	
	• Closure and remediation of existing unsanitary municipal landfills and open dumpsites and construction of sanitary landfills	
	• Construction of waste treatment facilities and recycling yards	
	• Construction of LFG collection system and leachate collection system and treatment	

3.12 UAE

Basem Abu Sneineh is the Director of General Waste – Waste Process & Treatment at Bee'ah. Basem has gained extensive waste management experience as an employee of Bee'ah since 2009. He oversees and manages the facilities at Bee'ah's waste management complex in Al Saja's, including a material recovery facility (MRF), a commercial and industrial (C&I) waste recovery facility, and an organic treatment facility. He is also responsible for alternative fuel, refuse-derived fuel (RDF), and solid recovered fuel (SRF) production facilities. Basem was

elected as the ISWA International Waste Manager-International Status from the International Solid Waste Association, Austria, in February 2018; was accredited by the Society for the Environment, UK, as a chartered environmentalist in January 2014; and was admitted as a corporate member by the Chartered Institution of Wastes Management, UK, in December 2013. He is also a member of the International Solid Waste Association (ISWA), Austria and the Institute of Industrial Engineers, USA.

3.12.1 Introduction

The UAE was one of the first countries in the Middle East to invest a large amount of money in waste management. It completed its largest environmental management project in 2007 under the name of Bee'ah, the Sharjah Environment Company. The responsibilities of Bee'ah include collection, processing, and recycling of all of the generated wastes in the Emirate of Sharjah, and it stands as the Middle East's fastest growing environmental management company. Bee'ah has four main holding companies:

1. Tandeef: The waste collection and management division (also in charge of public cleansing).
2. TadweerL: Waste sorting, treatment, and disposal.
3. Environmental Consultancy and Services.
4. Waste-to-Energy.

UAE has seven emirates and generates a total of 35 million tons of solid waste annually, of which 3 million tons are generated in the Sharjah emirate.

3.12.2 Collection

Collection, as mentioned before, is done by Tandeef, the waste collection, and management division (Figure 3.49, Figure 3.50 and Figure 3.51) Most of the waste in Sharjah (i.e., 95%) is collected as mixed waste and only a small portion (i.e., 5%) is source separated. Collection of domestic waste is done twice a day from every collection point in the highly populated areas and once a day in the less populated areas. The collection teams are highly trained to monitor and collect the waste not only from each collection point but also from any place along their collection routes.

Waste separation at the source has always been an essential concern for Bee'ah; therefore, a comprehensive transformation plan was started few years ago with the goal of separating 100% of Sharjah's waste at the source. The plan is currently in its preliminary execution stage. Many efforts have been made to increase the number of multicompartment public recycling bins and spread environmental awareness by educating, enlightening, and encouraging citizens to assimilate environmental values and take steps toward conservation and protection. With these steps and many more, it will not be long before Sharjah becomes the first city in the Middle East to separate all of its waste at the source.

Figure 3.49 Collection vehicles and workers.

Figure 3.50 Primary collection.

Figure 3.51 Secondary collection.

Tandeef is also credited with finding smart solutions for integrated waste management and utilizing the latest technology like geo-tagged smart bin sensors and automated route optimization. It has also an advanced fleet of eco-friendly vehicles for optimum performance, most of which are skip loaders, compactors, chain skip loaders, and dustcarts.

3.12.3 Processing and Recycling

Different types of waste are taken to different facilities for processing, as listed later:

- Domestic Waste: MRF (Figure 3.52).
- Construction and Demolition Waste: Construction and demolition recycling facility (Figure 3.53).
- Tire Waste: Tire recycling facility.
- Green Waste: Compost facility.
- Car and Metal Waste: Car and metal shredding and recycling facility.
- Commercial and Industrial Waste: Currently under construction.

Bee'ah has the world's third largest MRF, which is also the largest facility in the Middle East in terms of area and capacity of waste processed per hour. All of Sharjah's domestic waste is delivered to an MRF for processing. Wastes such as plastics, fibers,

Figure 3.52 Material recovery facility.

Figure 3.53 Construction and demolition recycling facility.

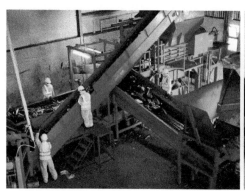

Figure 3.54 Biomass.

and metals are recovered in the MRF and sold to the customers, while the organic waste is transferred to the compost plant for further processing. The remaining unrecovered domestic waste (i.e. residues) is finally transferred to the engineered landfill (Figure 3.54).

3.12.4 Final Disposal

All of the recovery and recycling facilities, as well as the landfill (Figure 3.55), are located in a single complex called the Bee'ah Waste Management Complex that is in charge of managing all of the waste produced by the Emirate of Sharjah. There are no open dumps in the Emirate of Sharjah, nor in any of the UAE.

Bee'ah has a state-of-the-art engineered landfill that conforms to the highest international standards. It consists of a lined bottom, a leachate collection and treatment system, groundwater monitoring, gas extraction, and a cap system. As environmental protection is the highest priority for Bee'ah, leachate management is especially important, to ensure that the groundwater is not polluted, either in the short or long term.

Figure 3.55 Landfill.

By 2021, Bee'ah will finalize building its first waste-to-energy project in the Middle East. All the residues remaining from MRF recovery will be diverted from the landfill to the waste-to-energy facility, enabling them to reach their goal of "Zero Waste to Landfill."

3.12.5 Major Problems

- The lack of waste separation at the source is one of the major problems of waste management in Sharjah that also increases the burden on Bee'ah. This is considered the biggest challenge for Bee'ah because the Emirates of Sharjah is highly populated, with inhabitants from many nationalities (180) and different cultures. This makes spreading awareness and educating residents a complicated and costly mission.

- The items listed later should be considered for fixing the problems associated with solid waste management in Sharjah.

- A successful solution should be comprehensive, as there is no effective way to solve this problem except by involving all of the waste producers (i.e. the whole society) in the solution.

- Educating Sharjah residents is the first step, and it can be done at a faster pace by different means. Example of conventional means is increasing the number of campaigns and workshops.

- A reward system can be established for the residents and workers in the Emirate of Sharjah so that they are motivated by success, i.e. when a challenge is met as a result of their collaboration. The economic benefit that will be generated from having a higher recovery rate (and hence more recyclable materials) can be shared with Sharjah residents by reducing the waste collection tariff from each house's monthly bill. Special awards can be given, for example, to the best ecofriendly houses that comply with waste separation and/or for creative ideas that can enhance the application of strong waste management principles.

- Modern technology, such as smartphone applications, should be employed for protecting the environment. This will solve the waste separation-at-source problem and provide countless more benefits.

- Spreading awareness, introducing competitions, and keeping users updated and in direct contact with the latest environmental news and waste management technologies are some examples of things that excite most people.

Meanwhile, establishing more legislation for civil and industrial areas will remain a priority for solving the large-scale waste management problem that affects every resident of the Emirate of Sharjah. Thus, in the near future, Bee'ah can reach its goal of "Zero Waste to Landfill."

The summary of waste management system in UAE is presented in Table 3.17

Table 3.17 Waste Management in UAE.

Population	Country (UAE)	Major city (Sharjah)
Total population	9.68 million	1.63 million
Urban population (%)	93%	100%
Rural population (%)	7%	0%
Waste generation		
Waste generation/Year	35 million (6.5 million municipal)	3 Million (0.5 million municipal)
Waste generation/ Person/Year	2.4	1.84
% organic waste	11.00%	9.95%
Collection - % Collected	100	100
Type of vehicle and collection frequency	Skip loaders, compactors, and chain skip loaders Frequency: twice a day	
% mixed waste or % source separated	100% Mixed	95% Mixed 5.0% Source separated
Processing		
Material recovery facility and recycling (%)	21	66
Composting	Compost plant(s)	Compost plant
Disposal – Open dump		
Disposal – Engineered/ Sanitary landfill	N/A	Bee'ah Landfill
Landfill gas – Collection	N/A	Available
Landfill gas – Flaring or LFG	N/A	Available
Waste to energy	Dubai and Sharjah Waste-to-Energy Plants	Sharjah Waste-to-Energy Plant (Emirates Waste-to-Energy Company)
Major problems	Lack of source segregation	Lack of source segregation
Future needs	Source segregation, tipping fee implementation, and waste processing facilities	Source segregation and tipping fee implementation

3.13 VIETNAM

Tran Thi Diem Phuc *has a master's degree in environmental management from the National University of Singapore. She is the cofounder of Green Fingers Vietnam (2016), a youth-led nonprofit organization, focusing on educating children and raising communities' awareness about recycling and environmental protections. Her research interests center around sustainable solid waste management and the integration of the UN's Sustainable Development Goals within manufacturing models toward a circular economy. She hopes to contribute to the environmental protection movement step-by-step.*

3.13.1 Introduction

Vietnam is a country in Southeast Asia with an estimated population of 95.5 million inhabitants as of 2019. Its capital city, Hanoi, had a population of 7.8 million as of 2018. Most of the people in Vietnam live in rural areas (65.6%), while 34.4% reside in urban areas. Vietnam generates about 71.6 million tons (157 billion lb) of waste per year, of which Hanoi generates approximately 3 million tons (6.6 billion lb). The average per capita waste generation of the country is 273.75 kg (602.2 lb) annually. Overall, the organic matter in the MSW of Vietnam is 50-80%; however, the MSW of Hanoi is constituted of about 52% of organic waste. Figure 3.56 details the waste composition of Hanoi.

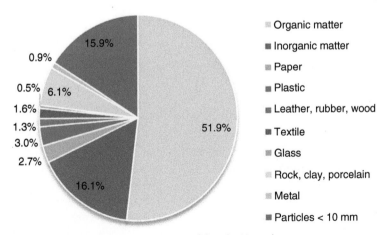

Figure 3.56 Solid waste composition in Hanoi.

3.13.2 Collection

The waste collection in Vietnam is 85% in urban areas and 40% in rural areas, while 92% of the waste is collected in the capital city The primary door-to-door collection in urban areas is carried out by a wheeled bin system, where the waste is collected from households and disposed of into containers of different sizes at selected locations in residential areas. Collection trucks are used for the secondary collection (Figure 3.57). Trucks collect the disposed of waste from containers and, depending on the capacity of their vehicles, go either to the transfer stations (small trucks) or directly to landfill (Figure 3.58) or treatment facilities (large trucks). The rural areas only have the collection

(a) (b)

Figure 3.57 (a) Waste collection truck and workers in rural area. (b) Waste collection truck in the city.

Figure 3.58 Waste pickers at a dumpsite.

truck and container system. The frequency of collection is once a day in big cities and two or three times a week in rural areas. Approximately 10-15% of the collected waste is source-separated in Vietnam.

In Hanoi, a wheeled bin/pushcart system is used for primary collection at least once per day, and street sweepers clean the major roads several times per day. Private companies with official licenses for waste collection and treatment are more common in large cities of Vietnam such as Hanoi and Hochiminh.

3.13.3 Processing and Recycling

Only about 10% of solid waste is recovered and recycled in Vietnam. The composting rate is a mere 4%. There are 14 waste-to-energy plants; however, they are used for waste incineration only, not for energy generation. The lack of proper source separation has hindered the recycling and waste processing industry and further separation of waste at treatment plants has created a burden as more workers are needed for the separation. Recycling is mostly handled by informal private sectors such as waste pickers, waste collectors, scavengers, and few small-scaled family businesses or enterprises. The informal recycling takes place both at the collection points and the dumpsites. There are many craft villages specializing in making products out of recycled plastic, paper, or metal. These villages have small family-operated businesses that produce a variety of products from recycled materials.

3.13.4 Final Disposal

Disposal in open dumps is still the most common method used for final disposal of solid waste. Of the 43% of MSW that is disposed of in open dumps (Figure 3.59), only 20% is taken to engineered/sanitary landfills (Figure 3.60) Open burning is a common practice in a few cities to lower the waste volume and increase the capacity of dumpsites, but it is more common in the countryside and rural areas. Due to the unavailability of proper waste incinerators, hazardous wastes from hospitals are dumped into the landfills.

3.13.5 Major Problems

The major problems in the solid waste management sector of Vietnam are listed later.

- The average salary in Vietnam is about VND 4 845 000/mo, and the fees paid for the collection service are very low (less than 0.5% of spendable income versus the international rate of 1–1.5%). Therefore, waste management services have to be subsidized by the government, which may affect the privatization of the sector.

- Only 30% of the waste disposal sites throughout the country can be classified as engineered landfills. The rest are open dumps, which contribute to bad odors, open burning of waste inside the landfills, and the release of leachate that pollutes the surrounding areas.

Figure 3.59 Working face of a dumpsite.

Figure 3.60 Working face of a landfill.

- Due to a lack of funding, most landfills in Vietnam do not have compactors, gas collection devices or facilities, leachate treatment, or environmental monitoring systems. Moreover, they are poorly managed.

- Leachate from the landfills contaminates groundwater and impacts the water wells of the communities living around the landfills.

- The discharge of toxic wastewater that has not been treated and/or poor operational practices contaminates the surface water.

- The vehicles used for collection are outdated.

- Burning from landfills/open dumps releases toxic gasses and causes health risks for the surrounding communities and scavengers.

- The open dumps attract animals (flies, cockroaches, and rats) that cause illnesses.

- Mismanagement of waste at the community level is leading to illegal discharge of waste to the canals, lakes, and the paddy fields.

The specific problems seen in Hanoi city are as follows:

- Nearly all of the collected waste is transported to the Nam Son landfill for disposal. The landfill is highly overloaded, and there is an urgent need for additional capacity for disposal. However, it is difficult to find land, and due to the environmental impacts of the current landfill operations, there is significant public opposition to landfilling.
- Until two to three years ago, compost plants were operated at Cau Dien and Kieu Ky. However, the operation was stopped as the compost could not be sold due to its poor quality.
- Many of the collection trucks are outdated and must be replaced by new/ additional compaction trucks.
- Recyclables are largely processed in craft villages, without proper monitoring of operating practices. These activities lead to substantial pollution of air, water, and land, which are serious health hazards for the workers.
- Due to the lack of infrastructure such as transfer stations, the current landfills are located more than 40 km distance from the city.

These problems can be solved by undertaking the following measures:

- Establishing a modern, integrated, and sustainable solid waste management system at an affordable cost to the government and the population.
- Designing collection and transportation systems, including transfer stations, and waste management disposal (upgrading) or treatment. Both rural and urban populations should be educated about environmental impact assessments, public hearings, government approvals, and financial arrangements.
- Replacement of open dumps with sanitary landfills.
- There are substantial environmental problems at the transfer points in the residential neighborhoods. Therefore, there is a need for well-planned and properly designed and constructed transfer points, where pushcarts and containers can be placed and then emptied into the secondary collection/ transport trucks, which should be designed for easy removal of excess solid waste.
- There is a substantial need for modern transfer stations outside the Hanoi city center, where waste can be compacted and transferred to larger transport trucks in order to substantially reduce the transport costs.
- Upgrading the collection equipment, disposal infrastructures (landfills), and the accompanying operational costs to improve the service levels and standards of the solid waste management system so that they do not negatively impact the environment and human health.

The summary of waste management system in Vietnam is presented in Table 3.18

Table 3.18 Waste Management in Vietnam.

Population	Country (Vietnam)	Capital (Hanoi)
Total population	95 562 935 (in 2019)	7 809 641 (in 2018)
Urban population (%)	34.4 (in 2019)	54.9 (in 2018)
Rural population (%)	65.6 (in 2019)	45.1 (in 2018)
Waste generation		
Waste generation/Year	71 671 000 tons/yr (in 2018) (158 billion lb/yr)	3 149 723 tons/yr (in 2018) (6.9 billion lb/yr)
Waste generation/ Person/Year	273.75 kg/person/yr (in 2018) (0.6 million lb/person/yr)	401.5 kg/person/yr (in 2018) (0.9 million lb/person/yr)
% organic waste	50-80	51.9
Collection – % Collected	85 in urban areas and 40 in rural areas	92
Type of vehicle and collection frequency	Wheeled bin system, collection trucks, and the container system in urban areas. The truck and container system in rural areas. Frequency: once a day (urban areas) 2–3 times per week (rural areas)	Wheeled bin/pushcart system, direct collection trucks, and the container system Frequency: Waste collection by means of push-carts is carried out at least once per day and street sweepers clean the major roads several times per day
% mixed waste or % source separated	10–15 separated waste of the collected quantity	NA
Processing		
Material recovery facility and recycling (%)	10	7.8
Composting	4	0
Disposal – Open dump (%)	43	NA
Disposal – Engineered/ Sanitary landfill (%)	20	NA
Landfill gas – Collection	NA	NA
Landfill gas – Flaring or LFG	NA	NA

(Continued)

Table 3.18 (Continued)

Population	Country (Vietnam)	Capital (Hanoi)
Waste-to-energy	14 (incineration only, do not generate energy)	0
Major problems	• Waste management services are subsidized by the government, which may affect the privatization of the sector • Only 30% of waste disposal sites throughout the country can be classified as engineered landfills; the rest are open dumps. • Lack of funding and poorly managed landfills • Contamination of groundwater and surface water due to leachate and toxic wastewater • Outdated collection vehicles • Burning from landfills/open dumps releases toxic gas causing health risks for the surrounding communities and scavengers • The open dumps attract animals, causing illnesses • Mismanagement of waste at commune level is leading to illegal discharge of waste to the canals, lakes, and the paddy fields	• Need for a new disposal site • Significant public opposition to landfilling • Inefficient composting facilities due to poor quality of compost • Outdated collection vehicles • Poor monitoring of craft villages for operational practices • Due to lack of transfer stations, current landfills are located more than 40 km from the city

Table 3.18 (Continued)

Population	Country (Vietnam)	Capital (Hanoi)
Future needs	• Modern integrated and sustainable solid waste management system at affordable costs to the government and the population • Design of collection and transportation systems, including transfer stations, and waste management disposal (upgrading) or treatment, including environmental impact assessments and public hearings, government approvals, and financial arrangements in both rural and urban areas • Replace the open dumps with sanitary landfills	• Well-planned and properly designed and constructed transfer points • Substantial need for modern transfer stations outside the Hanoi city center • Upgrading the collection equipment and disposal infrastructure (landfills) and the accompanying operational costs necessary to improve the solid waste management system to service levels and standards that do not cause environmental and health impacts

3.14 SUMMARY

The following problems have been identified as the most predominant among the 13 countries that were selected to represent different regions of the world for the case studies: open dumpsites and illegal dumping, health concerns caused by water and air pollution, and environmental issues caused by the destruction of the natural habitats of animals/birds. Some of the common solutions recommended are upgrading open dumpsites to sanitary landfills, collecting LFGs, and diverting MSW to recycling and composting facilities.

The lack of source segregation is a common problem seen throughout the world. Source separation at the household level is highly suggested, with incentives for public participation, such as tokens that could be used to pay for property taxes. This problem is directly associated with a lack of public awareness of solid waste management practices. Basic training programs to facilitate widespread awareness may be achieved through educational institutions and government involvement. Social initiatives that involve private and public sectors are also encouraged.

Table 3.19 Common problems and recommended solutions.

	Common Problems	Recommended Solutions
1	Open dumping, waste collection, and related health issues	• Engineered landfill • Gas collection
		• Circular economy
2	Source segregation	• Source separation at household level
3	Policy making	• Strengthening legislation
		• Improving private and public sector involvement
4	Lack of public awareness	• Training programs through educational institutes

Government involvement in making policies that manage solid waste effectively is minimum in most countries. The responsibility falls on the local governments, but the funding available to implement effective solutions is minimal. As a solution to this prevalent problem, it is suggested that policymakers revise and implement legislation that promotes sustainable waste management practices and allows sufficient funding for programs related to solid waste management. Involvement of stakeholders, NGOs, municipalities, government officials and experts will strengthen legislation that will help solve many of the current problems.

The common problems and recommended solution are presented in Table 3.19

REFERENCES

Abebe, M.A., and legese Abitew, A. (2018). Repi Open Dump Site Environmental Health Problem on Workers and People around It.

Abedin, M.A. and Jahiruddin, M. (2015). Waste generation and management in Bangladesh: an overview. *Asian Journal of Medical and Biological Research* *1* (1): 114–120.

Agência Portuguesa do Ambiente (2021). https://apambiente.pt/index.php?ref=16&subref=84&sub2ref=933&sub3ref=936 (accessed June 2021).

Ahmed, N. (2018). Karachi's trash. https://www.dawn.com/news/1382844.

Ahsan, A., Alamgir, M., Islam, R., and Chowdhury, K.H. (2005). Initiatives of non-governmental organizations in solid waste management at Khulna City. *Proceedings of 3rd Annual Paper Meet and International Conference on Civil Engineering, March* (Vol. 9, No. 11, pp. 185–196).

Ahsan, A., Alamgir, M., El-Sergany, M.M. et al. (2014). Assessment of municipal solid waste management system in a developing country. *Chinese Journal of Engineering* *2014* (12a): 1–11.

Alamgir, M. and Ahsan, A. (2007). Municipal solid waste and recovery potential: Bangladesh perspective. *Journal of Environmental Health Science and Engineering 4* (2): 67–76.

Ali, M. and Harper, M. (2004). *Sustainable Composting*. Loughborough University, UK: Water, Engineering and Development Center (WEDC).

Alvarez-Zeferino, J. C., Espinosa-Valdemar, R. M., Vázquez-Morillas, A. et al. (2017). *Producción de composta en la Ciudad de México: realidades y desafíos. VII Simposio Iberoamericano En Ingeniería de Residuos*, 297–302. https://redisa.unican.es/doc/actas-simposio.pdf.

Artelia (2013). Final Report – Solid Waste Management Strategy and Institutional Report.

Azevedo, A. L. M. D. S. (2021). IBGE – Educa | Jovens. IBGE Educa Jovens. https://educa.ibge.gov.br/jovens/conheca-o-brasil/populacao/18313-populacao-rural-e-urbana.html (accessed June 2021).

Banco Mundial (9 de Mayo de 2019). *Banco Mundial*. Obtenido de https://datos.bancomundial.org/indicador/SP.POP.TOTL?locations=CO.

Barua, V.B. (2016). Open dumps – Its effect on public life and health. *ISWA-SWIS Proceedings* 37–44.

BBS (2011). *Bangladesh Population and Housing Census*. Ministry of Planning, Bangladesh Bureau of Statistics.

BIGD (2015). *Solid Waste Management in Dhaka City: Towards Decentralized Governance, Sate of Cities 2015*. BRAC Institute of Governance and Development.

Biogas Doña Juana (Mayo de 2019). *Biogas Doña Juana*. Obtenido de http://biogas.com.co/#historia.

Bnamericas (Octubre de 2015). *BNAMERICAS*. Obtenido de https://www.bnamericas.com/en/news/waterandwaste/bogota-gets-fined-for-landfill-mismanagement.

C40 Cities (2016). C40 Good Practice Guides: Dhaka – Composting project. https://www.c40.org/case_studies/c40-good-practice-guides-dhaka-composting-project.

CCAC MSW Initiative (2014). *Climate and Clean Air Coalition MSW Initiative City Assessment*.

Central Administration of Statistics (2008). Lebanon in Figures, Republic of Lebanon.

Christian RL (2012). Feasibility Assessment Tool for Urban Anaerobic Digestion in Developing Countries. A participatory multi-criteria assessment from a sustainability perspective applied in Bahir Dar, Ethiopia. MSc thesis in Environmental Sciences (major in Environmental Technology), Wageningen University Netherlands, Final 24 January 2012.

Community Development Research (2011). *Ethiopia Solid Waste and Landfill: Country Profile and Action Plan*. Report produced with funding from the Global Methane Initiative.

Council for Development and Reconstruction (2009). Progress Report, Republic of Lebanon.

Council for Development and Reconstruction (2010). Work Program 2006–2009, Republic of Lebanon.

Council for Development and Reconstruction (2011). *Supervision of Greater Beirut Sanitary Landfills* (Contract No6823), Progress Report No 154.

Cruz-Salas, A.A., Alvarez-Zeferino, J.C., Vázquez-Morillas, A. et al. (2018). Composición de residuos plásticos en tres zonas del país. *Los Residuos Como Recurso 11* (1): 22–30. http://www.somers-ac.org/encuentros/encuentros.html.

Department of Environment (DOE) (2004). Country paper: Bangladesh. *Presented in the SAARC Workshop on Solid Waste Management*, held at Dhaka during 10–12 October 2004.

El Tiempo (9 de Augusto de 2018). *El Tiempo*. Obtenido de https://www.eltiempo.com/bogota/en-cuanto-quedaran-las-tarifas-de-aseo-en-bogota-para-el-2018-254004.

Enayetullah, I. and Hashmi, Q.S.I. (2006). Community based solid waste management through public-private-community partnerships: experience of waste concern in Bangladesh. *3R Asia conference*, Tokyo, Japan.

Eurostat (2021). Municipal waste statistics – Statistics explained, February. https://ec.europa.eu/eurostat/statistics-explained/index.php?title=Municipal_waste_statistics.

Federal Democratic Republic of Ethiopia Population Census Commission (2008). (rep.). Summary and Statistical Report of the 2007 Population and Housing Census. https://www.ethiopianreview.com/pdf/001/Cen2007_firstdraft(1).pdf.

Gaceta Oficial del Distrito Federal (2015). Norma ambiental para el Distrito Federal NADF-024-AMBT-2013, que establece los criterios y esecificaciones técnicas bajo los cuales se deberá realizar la separación, clasificación, recolección selectiva y almacenamiento de los residuos del Distrito Federa. Mexico. http://data.sedema.cdmx.gob.mx/nadf24/images/infografias/NADF-024-AMBT-2013.pdf.

Global Business Network Program (2020). Partnership Ready Ethiopia: Recycling Sector. https://www.giz.de/en/downloads/GBN_SectorBrief_%C3%84thiopien-Recyling_E_WEB.

Gobierno del DF (2003). NOM-083-SEMARNAT-2003, especificaciones de protección ambiental para la selección del sitio, diseño, construcción, operación, monitoreo, clausura y obras complementarias de un sitio de disposición final de residuos sólidos urbanos y de manejo especial (2004).

Green Earth Recycling (n.d.). http://www.greenearthrecycling.com.

Hai, F.I. and Ali, M.A. (2005). A study on solid waste management system of Dhaka City Corporation: effect of composting and landfill location.

Hussain, M. and Mushtaq, M.M. (2014). Awareness about Hospital Wastes and its effects on the Health of Patients in District Dera Ghazi Khan. *Asian Journal of Applied Science and Engineering 3* (3): 301–308.

IBGE (2018). https://cidades.ibge.gov.br/brasil/sp/sao-paulo/panorama.

INEGI – Instituto Nacional de Estadística y Geografía (2012). Medio ambiente: Residuos sólidos - tratamiento. https://www.inegi.org.mx/temas/residuos/default.html#Tabulados (accessed 5 May 2019).

INEGI – Instituto Nacional de Estadística y Geografía (2014). Medio ambiente: Residuos sólidos - recolección. https://www.inegi.org.mx/temas/residuos/default. html#Tabulados (accessed 5 May 2019).

INEGI – Instituto Nacional de Estadística y Geografía (2015). México en Cifras: Ciudad de México (09). https://www.inegi.org.mx/app/areasgeograficas/?ag=09 (accessed 4 May 2019).

INEGI – Instituto Nacional de Estadística y Geografía (n.d.). División territorial. http:// cuentame.inegi.org.mx/territorio/division/default.aspx?tema=T (accessed 10 September 2018).

Instituto Nacional De Estatistica (2021). *Special Highlight Statistics Portugal COVID-19* https://www.ine.pt/xportal/xmain?xpgid=ine_main&xpid=INE

ISWA Lebanon (2019). *Solid Waste: Lebanon an Overview*. Bilbao, Basque Country: ISWA General Assembly.

JICA. (2018). Future Vision of Solid Waste Management in Dhaka South City. Master Plan. Bangladesh: Dhaka South City Corporation, (draft).

Korai, M.S., Mahar, R.B., and Uqaili, M.A. (2017). The feasibility of municipal solid waste for energy generation and its existing management practices in Pakistan. *Renewable and Sustainable Energy Reviews 72*: 338–353.

Lisboa a Compostar (2021). *Câmara Municipal de Lisboa – Lisboa a Compostar | Sobre o Projeto*. https://lisboaacompostar.cm-lisboa.pt/pls/OKUL/f?p=178:15:888048765289:: NO: (accessed June 2021).

Maasri, R. (2012). Technical and economic measures for the rehabilitation and closure of solid waste dumps in Lebanon.

Mahar, A., Malik, R.N., Qadir, A. et al. (2007). Review and analysis of current solid waste management situation in urban areas of Pakistan. *Proceedings of the international conference on sustainable solid waste management* (Vol. 8, p. 36).

Ministerio de Vivienda, Ciudad y Territorio, Colombia (2015). *Minivivienda*. Obtenido de http://www.ciudadlimpia.com.co/site/images/Legislacion/20170509_1615_ DECRETO%201077%20DEL%2026%20DE%20MAYO%20DE%202015.pdf.

Ministry of Environment (2010). *Lebanon Country Environmental Analysis on Municipal Solid Waste Management*. Republic of Lebanon Draft Version No. 7.

Ministry of Environment (2011). *State of the Environmental Legislation Development and Application System in Lebanon*. University of Balamand, ELARD.

Ministry of Environment (2013). *National Implementation Plan for the Phase-Out of POPs in Lebanon*. Prepared by ELARD and ECODIT.

Ministry of Environment (2017). *Integrated Municipal Solid Waste Management Plan for Lebanon*.

ONU Medio Ambiente (2018). *Perspectiva de la gestión de residuos en América Latina y el Caribe*. Panamá: Ciudad de Panamá http://wedocs.unep.org/handle/20.500.11822/26448.

Pakistan – Waste Management (2018). https://www.export.gov/article?id=Pakistan-Waste-Management.

Pakistan Bureau of Statistics (PBS) (2017). *Population Census.* http://www.pbs.gov.pk/content/population-census.

Pakistan Environmental Protection Department (PakEPA) (2005). Guideline for Solid Waste Management. http://www.environment.gov.pk/images/provincialsepasguidelines/SWMGLinesDraft.pdf.

PCBA (2007). *Pollution Control Board Assam-Marches Ahead.* Guwahati, Assam: Pollution Control Board.

PETSTAR (2019). PetStar – procesos. http://www.petstar.mx/petstar/procesos (accessed 6 May 2019).

Plano De Gestão Integrada De Resíduos Sólidos da Cidade De São Paulo (2014). https://www.prefeitura.sp.gov.br/cidade/secretarias/upload/servicos/arquivos/PGIRS-2014.pdf

Plano Nacional De Resíduos Sólidos (2012). Governo Federal Ministério Do Meio Ambiente. https://sinir.gov.br/images/sinir/Arquivos_diversos_do_portal/PNRS_Revisao_Decreto_280812.pdf

Policy and Regulations on Solid Waste Management – Pakistan (2010). http://www.unep.or.jp/Ietc/GPWM/data/T2/AB_4_P_PolicyAndReg_Pakistan.pdf.

PONDUS Consulting (2009). Study on the possibilities of treating the waste from the Beirut slaughterhouse.

Relatório do Estado do Ambiente (2021). Municipal waste production and management. REA State of Environment Portal- Portugal. https://rea.apambiente.pt/content/municipal-waste-production-and-management?language=en (accessed June 2021)

Rizvi, J (n.d.). Lahore Cleans up Its Act. http://www.technologyreview.pk/lahore-cleans-act.

Sánchez-Velasco, E.L., van der Wal-Lima, M., del, M. et al. (2016). Operación de siete plantas de composta en la Ciudad de México. *Los Residuos Sólidos Como Fuente de Materiales y Energía 9* (1): 244–252. http://www.somers-ac.org/encuentros/encuentros.html.

Sarmiento, M.C. (2017). *Solid Waste Management in Colombia.* Dwaste.

SEDEMA – Secretaría del Medio Ambiente (2018). *Inventario de Residuos Sólidos 2017.* Ciudad de México: https://www.sedema.cdmx.gob.mx/programas/programa/residuos-solidos.

SEMARNAT – Secretaría de Medio Ambiente y Recursos Naturales (2015). *Informe de la situación del medio ambiente en México,* Capítulo 7 Residuos. México: Ciudad de México http://apps1.semarnat.gob.mx/dgeia/informe15/tema/cap7.html#tema1.

SEMARNAT – Secretaría de Medio Ambiente y Recursos Naturales (2016). *Residuos sólidos urbanos: la otra cara de la basura.* Ciudad de México: https://www.gob.mx/semarnat/galerias/residuos-solidos-urbanos-la-otra-cara-de-la-basura-18815.

Serbian Environmental Protection Agency (SEPA) (2017). Waste Management in Republic of Serbia from 2011-2017.

SNIS (2017). http://www.snis.gov.br/diagnostico-residuos-solidos/diagnostico-rs-2017 (data from 2017).

SREDA – Sustainable and Renewable Energy Development Authority. 2015. www.sreda.gov.bd.

SSPD (2017). *Informe de Disposición Final de Residuos Sólidos – 2017.* Colombia: Superintendencia de Servivios Públicos Domiciliarios.

Superintendente de Servicios Públicos Domiciliarios (2018). *Caracterización de organizaciones de recicladores de oficio en proceso de formalización.* Colombia: Superintendencia de Servivios Públicos Domiciliarios.

SWEEP NET (2014). Country report on the solid waste management in Lebanon.

Tassie, K., Endalew, B., and Mulugeta, A. (2019). Composition, generation and management method of municipal solid waste in Addis Ababa City, Central Ethiopia: a review. *Asian Journal of Environment and Ecology* 9: 1–19.

UN Environment (2019). *Waste Management Outlook for Latin America and The Caribbean.* UN Environment.

United Nations Environment Programme. (2007). (rep.). Lebanon Post-Conflict Environmental Assessment. https://postconflict.unep.ch/publications/UNEP_Lebanon.pdf.

UNEP (2010a). Assessment of the Solid Waste Management System in Bahir Dar Town and the Gaps identified for the Development of an ISWM Plan. Forum for Environment, June 2010.

UNEP (2010b). *Solid Waste Characterization and Quantification of Bahir Dar City for the Development of an ISWM Plan.* Forum for Environment, June 2010.

United Nation of Environmental Programme. (2010c). Solid Waste Management: Country Report. Lebanese Ministry of Environment.

UNEP (2015). *Global Waste Management Outlook.* UNEP.

UNEP (2019). *Waste-to-Energy: Considerations for Informed Decision-Making.* UNEP.

Waste Concern (2014). Bangladesh waste database 2014. Dhaka: Waste Concern (accessed 29 June 2019).

Waste Management World (WMW) (2015). *Recycling and Waste to Energy Project for Rawalpindi, Pakistan.* https://waste-management-world.com/a/recycling-and-waste-to-energy-project-for-rawalpindi-pakistan.

World Bank (2010). Draft Country Environmental Analysis Lebanon.

World Bank (2011). *PCB Inventory Update and Planning.* DRAFT FINAL REPORT.

World Bank (2017). *Data bank: Mexico.* https://data.worldbank.org/country/mexico (accessed 4 May 2019).

World Bank (2019). *Lebanon Population.*

Worldometer. (2020). *Lebanon Demographics.* https://www.worldometers.info/demographics/lebanon-demographics/

Chapter 4
Future Directions

Based on the current state of global waste management practices and the case studies of developing countries, it is very clear that a comprehensive and sustainable waste management system is needed. The proposed system should address the current problems and possibly convert the current problems into potential opportunities. Landfilling or open dumping has been the primary method of waste management globally for many years. In the last 40–50 years, developed countries have moved from open dumps to sanitary or engineered landfills, while

The Waste Crisis: Roadmap for Sustainable Waste Management in Developing Countries,
First Edition. Sahadat Hossain, H. James Law and Araya Asfaw.
© 2022 John Wiley & Sons Ltd. Published 2022 by John Wiley & Sons Ltd.

the developing world is still practicing the open dump method. Very recently the developing countries in South Asia, Latin America, Africa, and Eastern Europe have begun transitioning from open dumps to sanitary landfilling systems. However, based on the increasing rate of urbanization and the rate of increase in waste generation, the current ways of managing waste through landfilling or sanitary landfill systems may not be sustainable. Because of urbanization, it is almost impossible to find land on which to build a new sanitary landfill every 20–25 years, so that is not a sustainable solution. The problems associated with the waste generation and management by current systems are summarized below.

- Populations in urban mega-cities are increasing at an alarming rate, and the amount of waste that is being generated is also increasing.
- The people's purchasing power and living standard are increasing in many of the developing economies, and the amount of waste generation is increasing along with it.
- The available land for building residential houses to accommodate the influx of people from rural areas to urban areas is shrinking every day. Consequently, the cost of land for building a landfill within the city or in nearby cities is extremely high.
- The level of resistance to building new waste management facilities (landfills) within communities is very high, which causes

major distress within the cities' political system. Therefore, the policymakers or city officials are reluctant to approve the construction of new landfills within the city limits.

Moving from an open dump system to a sanitary landfill system may be a sound practice environmentally, but it is not a sustainable solution. Therefore, the proposed new concept of sustainable waste/material/resource management, which may or may not include biocells or perpetual landfills, is the most advanced solid waste management practice that has been proposed to date. A flow diagram of the proposed system is presented below (Figure 4.1).

The proposed sustainable **Resource Management** system would replace traditional landfills and open dumps and be a viable solution for developed and developing countries that need to eliminate the problems associated with them, such as loss of materials, climate change impacts, and post-closure monitoring costs. It will facilitate greater material recovery and reuse, accelerate waste degradation rates and renewable energy generation, perpetuate operations in the same location, improve the public's perception (by minimizing environmental impacts of open dumps), and enhance the acceptance by the greater urban community (by eliminating informal waste picker jobs that are dangerous and creating skilled worker positions). Developing countries have an excellent opportunity to "leapfrog" directly from untreated dumping to sustainable landfills (e.g. going directly from Stage A/B to Stage D, as seen in Figure 4.1), rather than progressing through open dumps and traditional landfills, as developed countries have typically done. The proposed sustainable resource/waste management system has three main components (Figure 4.2):

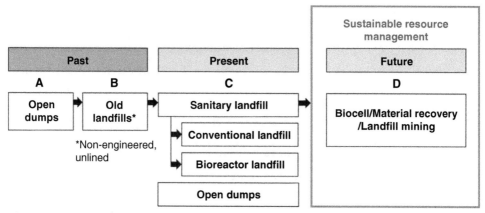

Figure 4.1 Waste/Landfill management in the past, present, and future.

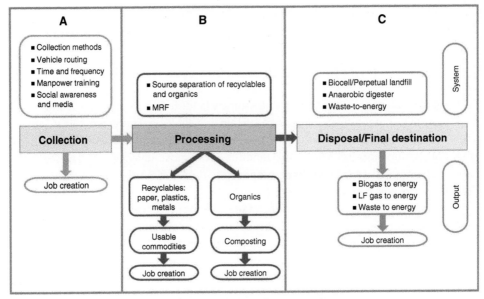

Figure 4.2 Sustainable waste management framework.

1. Increase waste collection in developing countries.
2. Process collected waste in material recovery facilities (MRFs) and increase the rate of recycling and number of ways to reuse recycling materials and/or create a sustainable market for recycled materials
3. Design the waste management facility as a resource recovery and management facility which may include a biocell, landfill mining, composting, landfill gas-to-energy or waste-to-energy.

This section focuses primarily on developing countries. The components of a sustainable waste management framework are presented below, and the sections are subdivided as follows:

a. Collection: Increasing waste collection through social awareness campaigns and encouraging entrepreneurs to participate in the waste collection business.
b. Processing of Waste: Source separation or MRF, reusing recycling products, creating jobs through manual and semi-automated processing at MRF, and providing and encouraging participation in training for social entrepreneurship.
c. Disposal/Final Destination: Perpetual landfilling through biocell landfill mining and landfill gas-to-energy or WTE. Initiating small-scale AD for biogas-to-cooking-or-energy at the community or household level.

Systems B and C can be in separate locations or the same location, depending on the waste collection strategies. For example, if source separation of waste is possible,

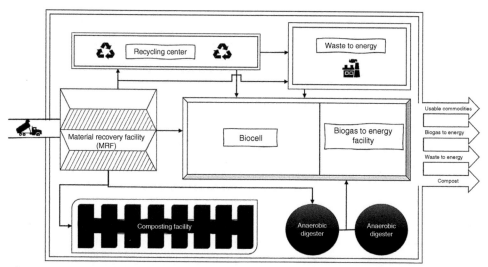

Figure 4.3 A SMART facility.

then the separated waste can go to different individual isolated places for processing and final disposal. However, since source separation is still a distant dream for many communities in developing countries around the world, we need to continue managing waste in a sustainable and healthy way while working on educating the public and creating awareness of the need for source separation. The concept of a **S**ustainable **M**aterial **A**nd **R**esource **T**reatment **(SMART)** facility (Figure 4.3) is presented here as a resource recovery and management facility. Once source-separated, mixed waste can be processed in a SMART facility, and sustainable material/resource recovery and management will be a ONE-STOP operation.

Landfills/dumpsites can be transformed into a sustainable waste management system that can be located in one place in perpetuity to generate renewable energy, compost, recycled materials, and biocells. The design and operation of conventional landfills will be replaced by a biocell, which acts as a perpetual landfill and is a major part of a sustainable waste management system in which waste is never landfilled permanently. Rather than serving as a permanent storage facility, the biocell is a temporary repository that is used to retrieve all of the potential benefits received during the repository period. It will also accelerate waste decomposition and replace landfilling by treating the waste so that it produces greater levels of renewable energy quickly.

Figure 4.4 presents a new concept of sustainable waste/material management and resource recovery that includes a perpetual landfill or biocell operation. A perpetual/ sustainable landfill (biocell) includes the up-front removal of plastics, glass, and metals. Since non-degradable plastics, glass, and metals are removed up front, all of the waste placed in the biocell can be degraded. This allows landfills to be in one location in perpetuity. Additionally, by accelerating and improving the overall degradation of

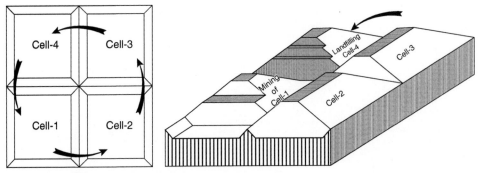

Figure 4.4 Biocell Operation.
Source: Based on Hossain et al. 2014.

waste, the biocell generates renewable energy, as organic waste completely degrades. Therefore, any materials that remain after degradation are organic in nature and can be reprocessed for use as compost, or at the landfill as an alternative daily cover material or as backfill soil during construction. The landfill is operated as a bioreactor landfill or biocell, with optimal moisture content being actively maintained to maximize waste degradation and methane generation rates. Referring to Figure 4.4, the biocells are advanced from 1 to 4. Since the waste will degrade quickly, it is expected that when waste is placed in Cell 4, the waste in Cell 1 will have already degraded and will be ready for reuse as compost or as the landfill's daily cover material. Mining would be conducted to retrieve the material, and the landfill space would be regained. The typical landfill life is 20 years, and each cell will be operated for five years. By the time that waste is placed in Cell 1 again, the waste in Cell 2 should have degraded and a mining operation can begin in Cell 2, which would be ready for re-use when Cell 1 is filled. In other words, there will always be about a 15-year period after a cell closes during which that cell can be mined. This will provide enough time for the waste to completely degrade and generate methane gas, potentially recoverable as biogas fuel, provided that the landfill has been efficiently operated as a biocell. The cells can continue to be re-used in perpetuity, eliminating the need to find a new site for additional landfill space. This would also eliminate the need for a final cover design and may extend the life of landfills more than 200 years (Figure 4.5) (Hossain et al. 2014).

Adopting this process means that waste management will move from an open loop system to a closed loop system to create a cyclic economy and enhance urban sustainability. The major benefit of the proposed method is to transform waste management to sustainable resource" management, which leads to better reuse of materials, higher production of renewable energy, decreased demand for landfill space, and improved public perception and acceptance by the greater urban community.

The proposed sustainable solution covers all of the major components of sustainability, as well as the technical, economic, and social aspects of the system. The government

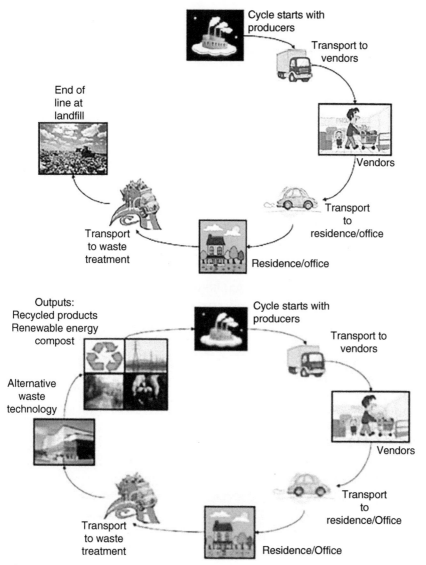

Figure 4.5 Open loop solid waste management to close loop solid waste management cycle.

Source: Based on Hossain et al. (2014)/with permission of American society of civil engineers.

shoulders the responsibility of collecting and managing waste. Consequently, a significant portion of a city's budget is allocated to municipal waste management program. The challenge is to shift the burden from the city to the private sector by creating an enabling environment for the business to flourish throughout the value chain from collection to processing waste to valuable products. These valuable products are then used by consumers while creating sustainable employment for urban dwellers with living income. Hence in the framework of Public Private Partnership (PPP), a SMART facility should be established to accommodate the interest of government, the private sector, and the public at large. Therefore, the SMART facility has the potential of creating jobs for the people living around the waste management facility, making it a source of income rather than a source of distress, disease, and economic hardship. If the waste management system is managed correctly and sustainably, the unhealthy conditions that are presently observed all over the world can be minimized significantly and thus provide healthy living conditions for both poor and rich residents.

4.1 MATERIAL FLOW IN SUSTAINABLE WASTE MANAGEMENT SYSTEM

A material recovery facility (MRF) can be built within the Sustainable Material and Resource Treatment (**SMART**) facility to remove the non-biodegradable plastics, glass, metals, inorganics, and even organic waste. An MRF processes a mixed solid waste stream and then proceeds to separate out designated recyclable materials through a combination of manual and mechanical sorting. Once the sorted recyclable materials are recovered from the waste stream, the organic waste can be separated from the mixed waste and diverted from final disposal through composting. If there is any leftover after composting and if it is available and cost effective for a country, it may be processed through an anaerobic digester. The balance of the mixed waste stream, if there is any remaining after recycling and composting, will be delivered for temporary disposal to the biocell or waste-to-energy facility, depending on the waste's characteristics. The materials' flow is presented in Figure 4.6. The degradation of the remaining wastes (without recyclables) is expected to be much faster than the degradation of solid waste mixed with non-degradable components; however, these remaining wastes can also be processed through a WTE facility, if technically feasible and if the community has available funding to build one. Only the leftovers go for temporary disposal in a biocell, which will be operated as a perpetual landfill.

4.2 PART A: SUSTAINABLE WASTE MANAGEMENT FRAMEWORK – WASTE COLLECTION

Based on the data presented in Chapter 2, the lack of collection of MSW is one of the main barriers to sustainable waste management. The waste collection rate in developing countries is so low that uncollected waste creates major environmental, social, health and economic problems. The failure to properly collect waste from the

(a)

Option 1: Mixed waste collection

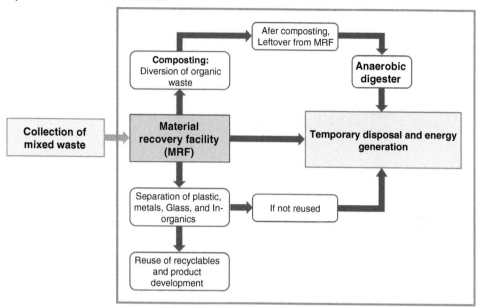

Option 2: Both mixed waste and source separated waste collection

(b)

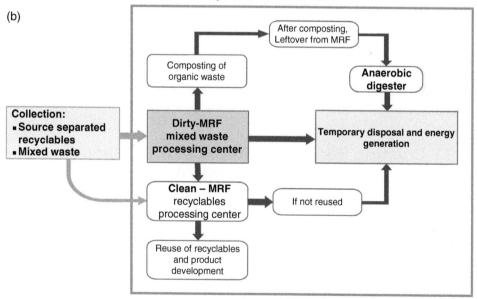

Figure 4.6 Material flow in sustainable resource management system (a) Mixed waste collection (b) Both mixed waste and source separated waste collection (c) Source separated waste collection.

Option 3: Source separated waste collection

(c)

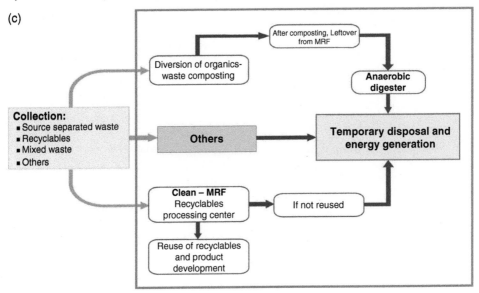

Figure 4.6 (Continued)

point of generation is almost as significant of a concern as the mismanagement of the materials at the point of disposal. For example, major metropolitan cities or mega cities are experiencing flash flooding during the rainy season because of uncollected waste clogged drain lines preventing proper flow of stormwater. Therefore, it is important for us to develop a methodology or ways to increase the efficiency of waste collection, mainly in developing countries.

4.2.1 Creating Social Awareness of Importance of Waste Collection and Management

In developed countries, it is the responsibility of City Officials to protect public health and the environment. This obligation is reflected in the development of programs to manage municipal wastes and the enforcement of environmental regulations. Residents in most undeveloped countries are not aware of the negative impacts of their trash on public health, the environment, and their own economic condition. The trash that they throw away is initially "out of sight, out of mind"; but it eventually is very much in their sight and mind because it accumulates in every corner of the city. The general view of almost everyone is that it is the public officers' or city regulators' job to clean up the mess that they have created. It is important for members of communities to understand that waste management is the responsibility of everyone, and the government officials as well as the citizens need to do their part. Without the active participation and engagement of every citizen in the city/locality/town/country, sustainable waste management will remain an unsustainable dream and will never materialize.

City officials and regulatory authorities need to initiate massive social campaigns to increase awareness of the importance of sustainable waste management. Active campaigns on all media outlets are desperately needed to convey the potential impacts on health and human life of uncollected or mismanaged waste, and interdisciplinary teams of environmental scientists, social experts, the media, and marketing experts should work together to create a successful campaign. In many countries, national and local TV outlets owned by the government or local agencies have been utilized extensively to reach out to the public. Moreover, due to the use of social media by all sectors of society, it is also possible to launch social media campaigns to create awareness. Some of the fundamental ideas are presented below.

1. TV and Newspapers

A slot should be created through different TV/newspaper outlets to promote the need for waste management during prime time. The government can obtain the slots at a cheaper rate and should do so, as it is being done for the general public and not for profit-making. Cartoons and TV shows can be created to reach children at an early age, as these will have a long-term lasting effect.

2. Social Media

Social media can be used to create awareness and increase collection by informing/alerting the residents on waste collection dates and pick-up locations. Residents can also take photographs of uncollected waste and send them to city officials with locations to collect those uncollected wastes. People are highly engaged in social media thereby creating awareness and touting the benefits of a sustainable waste collection system.

3. School-and-University-Level Curriculums and Competitions

In many countries in the world, topics related to trash/garbage are included in the primary/secondary education system. Creating an awareness among young children seems to have had very positive effects in the developed world, especially in the U.S. and Europe; therefore, it is important for developing countries to make long-term plans for increasing awareness, beginning with primary education. Competitions among high school and college students based on waste recycling, city cleaning, community volunteer services, can motivate them through different levels of awards to inspire, educate, and even create future waste management leaders.

4. Monthly Clean-Up Programs

Monthly neighborhood clean-up programs can energize both youth and adults and motivate them to participate in neighborhood cleaning programs. If the program is advertised ahead of time, it can bring together many people to clean up a neighborhood, city, shopping complex, or village market. There are examples of many such programs that are very successful if they are continued on a regular basis. Creating awareness and educating the public are not going to happen overnight and will require extensive planning, whole-hearted implementation, and above all, the intent of both the regulators and the residents for a healthy urban city development. In the meantime, the following things can be done when planning public awareness.

- Avoid throwing garbage on the road or in drains or water bodies, or any other places not designated for trash.
- Be mindful of the waste collection schedules and the instructions pertaining to how and where to place the waste.
- Source separate the waste.

The government/city officials/regulatory authorities should make sure to:

- Schedule collections on a regular basis.
- Provide residents with separate bins for dry waste and wet waste.
- Train employees to use advanced vehicles to collect more waste and more efficiently.
- Train employees to use different vehicles for collecting dry and wet waste from waste collection zones.

The major benefits of community involvement and social media campaign are: (i) a sense of belonging and community, (ii) group accomplishments and social involvement, (iii) making money while cleaning up the environment, (iv) creating jobs at many levels of waste management, and (v) having fun while helping the community and the environment.

4.2.2 Mixed Waste vs. Source Separated Waste

A sustainable waste management system demands the enforcement of source separation of solid waste, which is vital to better management of waste and may result in a higher quality of compost, the use of the separated organic/food waste for biogas generation through an anaerobic digester, and the creation of new products from the recycled materials. In some cases, when waste is placed in community bins, waste pickers sort through the waste piles and take the recyclable materials out of waste (informal sector) so that by the time the city trucks arrive to pick up the waste, most of the recyclables are gone. Another major issue in developing countries is that the food waste and organic waste (which can be up to 70% of the waste) are mixed with other waste, making it extremely difficult to use it for other applications like composting.

Source separation of solid waste is a problem in both developed and developing countries. In the U.S., many states and cities do not have residential recycling bins, so they only collect mixed waste (Figure 4.7a). Even for those who have the option of recycling, the regulations are very flexible, and the residents are not required to use recycling bins.

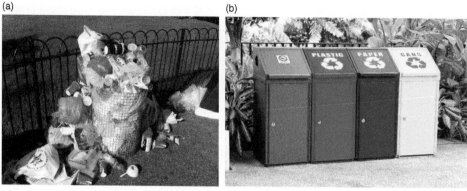

(a)   (b)

Figure 4.7 (a) Mixed waste bins in the US (https://www.planetaid.org/blog/trash-volumes-increase). *Source*: oatsy40/Flickr. (b) Recycling bins in Europe. *Source*: United States Environmental Protection Agency, 85. https://www.epa.gov/schools/appendix-model-program-state-school-environmental-health-guidelines.

In many European countries, however, households have multiple bins into which they separate their waste by category (food waste, organic waste, plastics, papers, metals, and others), and in some cases, source separation is mandatory at the household level (Figure 4.7b).

It is therefore no surprise that waste management is more advanced in many European countries, mainly in western and central Europe, than any other countries in the world. The eastern European countries, on the other hand, are far behind from the rest of the Europe and in some cases many developing countries, in managing their waste.

It is not easy to change from mixed waste to source-separated waste; Cities can develop educational program to create awareness among residents for the urgency of source separation and develop private public partnership (PPP) programs that allow private companies to participate in the collection program. For example, composting companies could provide households with bins that are designated only for degradable waste so that the collected organic waste is not contaminated. This would help the composting companies make better products and, at the same time, reduce the burden on the city, as the volume of waste would be significantly reduced.

4.2.3 Collection Vehicles

The choice or availability of collection vehicles is very important for increasing waste collection, but advanced waste collection vehicles are not available in most developing countries. Many of the developed countries have advanced waste collection trucks that are equipped with an automated compaction system which greatly enhances the efficiency of their operation by increasing the volume of waste that can be collected. These collection vehicles are rare in developing countries. If a city mandates source separation of solid waste, it needs to have two different trucks for waste collection: one for mixed waste and one for dry waste or recyclables. The old collection vehicles that are used by developing countries only have a very low collection volume compared to developed countries and the initial collection flow is time consuming. The following changes are recommended to increase the efficiency of collecting waste.

i. Better or advanced collection vehicles: Developing countries need to invest in better or advanced waste collection vehicles. It may be a difficult process, but with a good design and implementation plan, they can probably secure funding from the World Bank or other financial institutions. Additionally, the efficiencies gained with improved collection vehicles, as well as the safety to the workers would offset the cost of inefficient collection systems. This would be a landmark change in waste collection and would make great strides toward achieving sustainable waste management operations and cleaner cities in the long run.

ii. Private/Public Partnership: Cities can work with private companies that use waste as their business's raw material to collect waste from specific sections of the city, as this would reduce the size of the investment needed for purchasing waste collection vehicles. For example, vegetable market waste is purely organic and is valuable for composting companies. The city can assign specific companies to collect market waste directly.

iii. More community bins can be distributed across the city and residents can be required to deposit their waste into the bins on a specific day(s) of the week. This would minimize the initial lag phase and use of community vans and has the potential to increase the efficiency of the collection process while instilling a sense of community and pride among the residents.

4.2.4 Creation of Different Zonings for City Waste Collection

City officials can divide and subdivide the city into many different sections and zones by communities, businesses, and population densities. These sections and zones will be subdivided into many different subsections or subzones, and each subsection will be managed by a supervisory position, which will be created as part of the new project. The details of this zoning will be provided during the project implementation. According to the zoning, collection routing can also be designed in a way to reduce the collection time. Community level supervisors can act both as operators and implementors of policy. They can knock on doors if any specific houses are not following protocols and can also implement a fine policy for residents. Creation of this group can be done by the city or can be contracted to private companies for management. This will create many community level jobs and have a greater chance of success as every zone supervisor has their responsibility to keep up their section/zone cleanliness to maintain their job status.

4.2.5 Collection Time and Frequency

Collection times will depend on the size of the city, traffic flow conditions during rush hours, and the starting times of academic institutions (mainly elementary and secondary schools). Collection schedules should be designed so that they do not complicate the already heavy traffic volume of early morning rush hour traffic. Collection could begin in residential areas as early as 6:00 a.m. in non-urban areas and as early as 2:00 a.m. in commercial and urban residential areas. Waste collection is done in Portugal at night to avoid daytime traffic (Figure 4.8a, b).

4.2.6 Training and Creating Skilled Manpower

Highly skilled and trained manpower is a must for sustainable waste management systems in both urban and rural settings. Every city should ensure that their collection and waste management crews are well trained, either by enrolling them in local or international training facilities or by bringing in trainers/experts specifically for that purpose. A sustainable waste management system cannot be accomplished without a trained and skilled workforce, as they are at the core of increasing waste collection and creating a healthy and clean urban city.

(a)   (b)

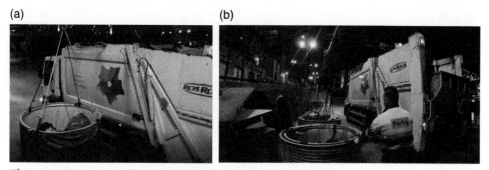

Figure 4.8 Waste collection at night to avoid traffic (a) Waste Collection Vehicle in Portugal; (b) Waste Collection Personnel working at Night.
Source: 86 Photo taken by Soraia Taipa (SWIS Ambassador).

4.3 PART B: SUSTAINABLE WASTE MANAGEMENT FRAMEWORK: WASTE PROCESSING AND RECYCLING

4.3.1 Material Recovery Facility (MRF)

A material recovery facility (MRF) is a processing center through which resource recovery and reuse can be implemented. Before waste is finally disposed. A mixed waste collection system with no source separation means that all of the resources for a circular economy will end up in landfills and the idea of resource recovery will be eliminated. Therefore, MRF is a very important and critical infrastructure for sustainable waste management.

An MRF accepts source-separated or mixed material (solid waste), separates the materials into individual waste types and prepares it as raw material for future use or further manufacturing. The main purpose of the MRF is to separate the recyclable material from the food waste or organic materials. A MRF can be operated by mechanical or manual separation techniques (Figure 4.9). Older MRFs rely heavily on manual sorting, which can lead to high quality material recovery, but it is expensive and time consuming in developed countries where the workers' hourly rates are high. Labor cost becomes the highest component of the operating cost for manual MRF in developed countries, but it can be relatively inexpensive in developing countries, where the workers' hourly rates are low compared to those in developed countries. This can also be a major source for jobs for the people living around the MRF (which can be co-located with landfills or waste management facilities).

(a)   (b)

Figure 4.9 (a, b) Manual sorting in MRF (https://www.ecomena.org/materials-recovery-facility; https://www.esri.com/about/newsroom/arcuser/addressing-locations-of-recycling-offenders/).
Source: Content is the intellectual property of Esri and is used herein with permission. Copyright © 2022 Esri and its licensors. All rights reserved; Ignácio Costa/Wikipedia Commons/Public Domain.

Automated sorting incurs higher equipment costs but results in greater material recovery and faster processing rates. Additionally, automation has the advantage of reducing health and occupational problems that can be caused by direct handling of the bulk waste. Conveyor belts transport waste materials throughout the facility to each area of separation associated with the mechanical equipment within the MRF. Most facilities have at least an initial sorting station with flat belt conveyers to provide easy access to the materials on the belts. These initial sorting areas are intended to remove the obvious objectionable materials.

There are two types of MRFs, clean and dirty, and the source-separated recyclable materials go to clean MRFs. A schematic diagram of a clean MRF is shown in Figure 4.10a. A dirty MRF accepts unsegregated bulk solid waste that requires manual sorting to separate the recyclables from the mixed waste. Figure 4.10b shows a schematic diagram of a dirty MRF. The inputs and processing mechanisms of clean and dirty MRF is summarized below (Table 4.1).

Recyclable materials collected from households (if source separated) or after sorting at material recovery facilities (MRFs) need to be used for product development to reduce the burden of utilizing virgin material. This also supports the idea of a circular economy, where

(a) (b)

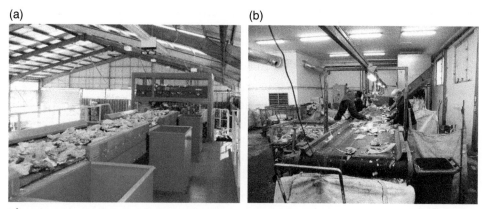

Figure 4.10 MRF Composting (a) Clean MRF. *Source*: https://www.game-engineering. com/products/material-recycling-facilities-mrfs. (b) Dirty MRF. *Source*: Michal Maňas/ Wikipedia Commons/Public Domain.

Table 4.1 Inputs and Processing mechanism of clean and dirty MRF.

Type	Inputs	Processing
Clean	Clean or source separated	Manual/Mechanized
Dirty	Dirty or non-source separated	Manual/Mechanized

Source: Department of Environment and Natural Resources (2003); Lopez and Kemper (2008).

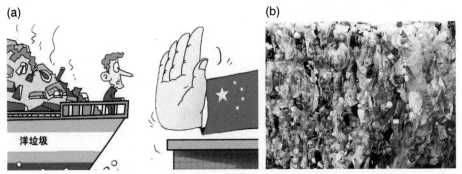

Figure 4.11 (a) China banned foreign waste from 1 January 2018 (CGTN).
Source: RTM World., https://www.rtmworld.com/news/china-to-ban-solid-waste-imports.
(b) Piling up: How China's ban on importing waste has stalled global recycling by Cheryl atz, 7 March 2019 (Yale Environment 360). *Source*: Hans / Pixabay.

reusable materials are not thrown into landfills or incinerated in a WTE plant. Recycling and reusing recycled products have been profitable businesses for many years and China was considered the global hub to export recyclable materials for the rest of the world. China had been the world's main importer of global waste for nearly 30 years. However, the enactment of the China Sword (China's deep restriction on importing recycled waste materials) in January 2018, changed everything overnight for the recycling industry (Figure 4.11).

4.3.1.1 Immediate Impact of China Ban

The world entered a new reality with the enactment of the China's Sword. Every country in the world had been sending their recyclables to China and consequently had not developed a local market for making and selling recyclable materials or products. The U.S. recycling facilities are now struggling to find markets for paper, cardboard, and plastic, the three staples of the recycling industry that continue to rise in usage (Figure 4.12).

Some of the immediate impacts of the China ban are listed below.

- Diminishing recycling program: The China Sword, which enacted significant limitations on "contamination" essentially functioned as a ban and has had a very serious negative impact on the recycling industry (Figure 4.13). Many cities across the world are struggling to continue with recycling practices as their market for recycling products has plummeted. Many cities have even stopped their recycling programs, as they lack a sustainable funding mechanism. Based on the available local data and waste characteristics in Texas, the percentage of plastic content in MSW going to landfills increased from 10–12% to 18–19% in 2019, one year after the China ban. That is an almost 50% increase in the plastic content of the MSW stream.

- Relocation of Chinese companies

- Import focus on other Asian countries

(a)   (b)

Figure 4.12 Recycling materials are piling up around the world (a) USA (https://
www.nwpg.org/2018/06/13/this-is-why-a-lot-of-our-recycling-is-going-to-landfills).
Source: Philip/Adobe Stock. (b) Adelaide, Australia (https://www/independent.
co.uk/climate-change/news.recycling-uk-cardboard-b1817870.html). *Source*:
JMacPherson/Flickr.

Figure 4.13 Following China
ban countries deal with burden
of plastic waste at Vietnam.
Source: pxhere.com.

Since the initial restrictions, China has now fully banned the import of recyclables. After the initial restrictions, many recyclables were diverted to other south Asian countries. Most of the other south Asian countries, Malaysia, and Vietnam have now followed China's example and banned the importation of recyclable materials. Since July 2018, Malaysian government officials have shut down at least 148 unlicensed plastic recycling factories because of their noncompliance with environmental policies and illegal burning of plastics. The effects of these unlicensed plastic recycling factories are:

- Open burning of recyclables is destroying large amounts of reusable materials.
- Mercury, dioxins, lead, carbon dioxide and other pollutants are emitted from burning waste and have very serious negative effects on human health.
- Burning of plastics produces toxic ash that has to be disposed of in landfills.

The South Asian bans are summed up with this news flash! "Malaysia sends back trash, says will not be world's waste bin." AP, 20 January 2020, Environment Minister Yeo Bee Yin says "Our position is very firm. We just want to send back (the waste), and we just want to give a message that Malaysia is not the dumping site of the world".

Recycling has declined over the last two years (Citylab 2019), and amid this crisis, opportunities are greater for waste and recycling businesses in developing than in developed countries, as the developing economies are paying closer attention to their waste management systems. Many of them are just beginning to initiate waste management practices, and the potential for young entrepreneurs to clean their villages, towns, cities, states, and countries and to make money from doing it is tremendous. They, however, do not have a market for recyclable goods, which currently go to landfills, and it is critical that they create local markets for utilizing these materials.

4.3.2 The Impact of COVID-19 on Plastic Waste

Plastics are highly nondegradable and can require as long as 500 years to completely degrade. Although they are lightweight, they occupy a large volume of landfill space and remain in the landfill forever, reducing the amount of space available for more environmentally friendly waste (Hossain et al. 2017). Therefore, the diversion and/or reuse of plastic products from landfills could save landfill space and help the landfill to stabilize at a faster rate. The COVID-19 pandemic brought many changes to the lives of residents and to businesses across the country. While some services were ordered to close because they were considered non-essential, some have been able to continue operating even during the lockdown period, as they were deemed essential. The collection and processing of municipal solid waste is among the businesses considered essential; therefore, their employees have continued to work throughout the lockdown and pandemic period. As the coronavirus can survive on solid surfaces for 24–72 hours, it is important to ensure that the health and safety of these workers are not at risk. The following are a few of the issues that needed to be addressed for their health and safety.

1. Generation and classification of COVID-19 waste.

2. Issues pertaining to the littering of COVID-19 waste and its consequences.

4.3.2.1 Generation and Classification of COVID-19 Waste

The quality of the air and water improved significantly globally during the coronavirus lockdown; however, the generation of medical and household waste increased significantly. During the peak of the epidemic in Wuhan, China, 240 tons (0.5 million lb) of medical waste were generated every day, which is six times more than what was generated before the pandemic (Zuo, South China Morning Post 2020) (Figure 4.14). In Wuhan alone, one new medical waste treatment facility and 46 mobile waste treatment facilities were built to accommodate the increased amount of waste. To complicate matters, the amount of medical waste related to COVID cannot be accurately predicted in many countries, as the pandemic has created serious havoc in many metropolitan cities in Europe and the USA.

The amount of municipal solid waste generated independently from COVID-19 also increased significantly. There are many reasons that this has occurred, but a few are listed below.

- During lockdown, more people stayed at home to eat because most of the restaurants were closed.

- Many cooked just to have something to do during lockdown.

- Kids stayed home and ate more snacks, or some cases, cooked with their parents to fight the boredom resulting from staying home all the time (Figure 4.15).

Figure 4.14 Recycling materials are piling up around the world including medical waste.
Source: DFID/Wikipedia Commons/ Public Domain.

(a)   (b)

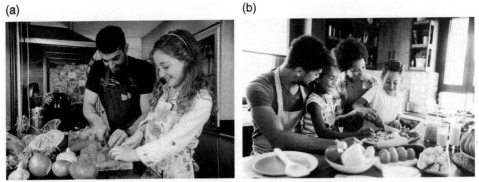

Figure 4.15 (a, b) Cooking and family time spent during the COVID 19 Lockdown (https://www.unicef.org/azerbaijan/stories/easy-affordable-and-healthy-eating-tips-during-coronavirus-disease-covid-19-outbreak; https://www.dailynews.com/2020/10/28/5-food-rituals-that-help-create-healthy-eating-habits-and-nutrition). *Source*: © UNICEF/UN0343152/Pazos; NDABCREATIVITY/Adobe Stock.

- Online shopping, including grocery shopping, skyrocketed, significantly increasing the amount of packaging, paper, and plastic waste. Based on data from different sources, residential waste generation increased up to 40% and commercial waste generation decreased by 20 to 30%. as most commercial facilities were closed. Overall, there has been an increase of about 10 to 15% in the overall volume of MSW.

On 26 March, Justine Calma of The Verge reported that COVID-19 waste can be categorized as (i) waste generated at hospitals, medical facilities, and healthcare facilities, or (ii) waste generated outside of hospitals, medical facilities, and healthcare facilities (Calma 2020).

1. COVID-19 waste generated at hospitals, medical facilities, and healthcare facilities (for example, masks, gloves, lab specimens, needles, cleaning cloths, wipes, single-use microfiber/plastic cloths, etc.) (Figure 4.16) can be treated the same way as regular medical waste, according to the Center for Disease Control (CDC). In United States, medical waste treatment varies by location and can be governed by the state, OSHA (Occupational Safety and Health Administration) or the Department of Transportation. In general, contaminated medical waste should be burned, sterilized, or chemically disinfected before being disposed of in a landfill.

On 13 March 2020, the Texas Commission on Environmental Quality (TCEQ) recommended following the CDC's and the World Health Organization's (WHO) recommendations for managing waste materials related to COVID-19 that originate from healthcare facilities. Not all of the trash related to COVID-19 comes from

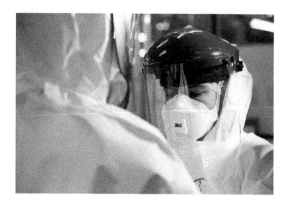

Figure 4.16 Health care workers don personal protective equipment. *Source*: Wikipedia Commons.

(a)   (b)

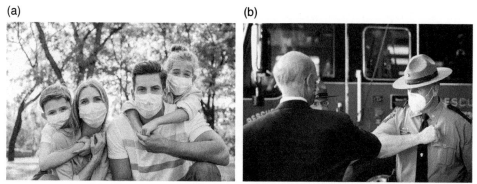

Figure 4.17 (a) Members of a family, all wearing masks, pose for a picture. *Source*: Stocksnap.io. (b) Police officers wear protective masks. *Source*: Governor Tom Wolf/Flickr.

healthcare facilities, as residents, the police, postal workers, and others working outside their homes wear masks and gloves that end up in the trash (Figure 4.17). Whether people are hospitalized with severe COVID-19 symptoms or recover at home from moderate-to-minor symptoms, their trash is contaminated (Figure 4.18). This is a bigger problem than was anticipated and can create health and safety issues for residents and waste collection and processing workers. The coronavirus can stay alive on cardboard, metal, and paper for 24 hours and on plastic surfaces up to 72 hours and can infect the waste workers and the general public if they come in contact with it within 72 hours. Both residents and waste workers need to pay close attention to their actions to ensure their health and safety.

The CDC, the Occupational Safety and Health Administration, and a few state agencies have taken the stance that businesses outside the realm of hospitals, medical

(a) (b)

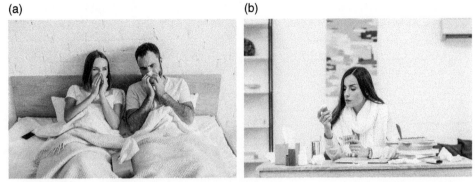

Figure 4.18 (a) COVID-19 symptoms vary person to person, as does the length of the infection. *Source*: LIGHTFIELD STUDIOS/Adobe Stock. (b) Experts say you should have groceries, medical supplies, and cleaning supplies, as well as other items on hand. *Source*: LIGHTFIELD STUDIOS/Adobe Stock.

facilities, and other healthcare entities should manage waste that is potentially or actually contaminated with COVID-19 in the same manner that they manage any other MSW. They espouse that such waste is not considered hazardous waste or regulated medical waste that needs to be specially handled, managed, and disposed of.

In an April 25 article, "The Amount of Plastic Waste is Surging Because of the Coronavirus Pandemic," that was published in Forbes, Laura Tenenbaum reported that there has been a tremendous increase in the volume of plastic waste generated during the pandemic, and it is creating another severe environmental problem. There are many reasons for the increase in plastic waste during lockdown:

i. Most PPE (masks, gloves, and other medical equipment) is made from plastic and is widely utilized during the pandemic (Figure 4.19). "The PPE is intended to help fight a public health challenge, not create a plastic pollution problem," Adrienne Esposito, executive director of Citizens Campaign for the Environment, told CNN (Sangal 2020). However, we need to pay attention to the amount of additional waste we are generating and determine whether how we are disposing of it is making PPE waste management sustainable.

ii. In an effort to reduce the amount of pollution from plastics and its effect on the environment, many U.S. cities and states have banned or are working on banning one-time-use plastic bags. Some members of the plastic industry are taking advantage of the situation, however, and are creating fear by claiming that it is safer to use the single-use bags. Laura Tenenbaum of Forbes reported that since the beginning of the coronavirus epidemic, many grocery stores have begun using single-use plastic bags and have discouraged or prohibited shoppers from using their own reusable bags (Tenenbaum 2020) (Figure 4.20).

(a)　　　　　　　　　　　　　　　　(b)

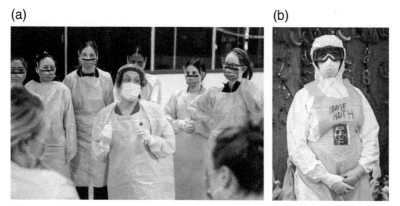

Figure 4.19 (a) Medical staff prepares to receive patients for coronavirus screening at a temporary assessment center at the Brewer hockey arena in Ottawa, Ontario. *Source*: PATRICK DOYLE/REUTERS/Adobe Stock. (b) Healthcare workers stick smiling photos of themselves on PPE to reassure COVID-19 patients. *Source*: Mary Beth Heffernan / Wikipedia Commons / Public Domain.

(a)　　　　　　　　　　　　　　　　(b)

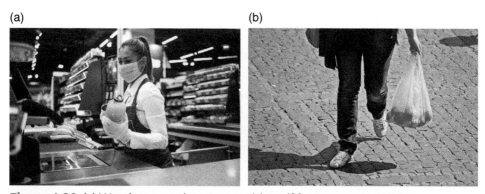

Figure 4.20 (a) Wearing a mask to protect himself from exposure to the coronavirus. *Source*: littlewolf1989/Adobe Stock. (b) Use of single use plastic bags for shopping. *Source*: cocoparisienne/Pixabay.

iii. In a report for CNN, "Coronavirus is Causing a Flurry of Plastic Waste," Rob Picheta said that COVID-19 has accelerated the production of desperately needed plastic products, as everyone wants their share to safeguard themselves against the virus (Figure 4.21) (Picheta 2020). "We know that plastic pollution is a global problem - it existed before the pandemic," Nick Mallos of the U.S. - based NGO Ocean Conservancy told CNN, "(But) we have seen a lot of industry efforts to roll back some of the great progress that's been made. We need to be quite cautious about where we go, post-pandemic."

(a) (b)

Figure 4.21 (a)/(b) The University of Texas at Arlington students wearing plastic over their clothes during waste sample collection for the City of Irving Landfill, Texas (Photo taken by SWIS Member).
Source: Courtesy of SWIS.

4.3.2.2 Issues Pertaining to the Littering of COVID-19 Waste and their Consequences

PPE has been widely used throughout the pandemic, and most of it is made of plastic. It is vital to the safety of the general public and frontline workers and will continue to play an important role in our safety as long as the coronavirus is a threat. Close attention needs to be paid to whether the additional amount of waste that is being generated is being disposed of in a safe and sustainable manner. Littering and illegal dumping of waste create eyesores, but more importantly, harms the environment (Figure 4.23). Microplastics that are thrown away in open public spaces by the public, who give little thought to the consequences of their actions, are being washed into waterways, and many of them are ending up in the ocean (Figure 4.24). Many are being added as people visiting beaches discard these materials on the beaches or in the water (Figure 4.22). Hannah Frishberg, in the NY Post on 21 April 2020, reported a similar concern in her report, "Littered Masks and Gloves Filling Streets, Becoming Safety Hazard." They are clogging sidewalk drains, and if enough of them end up in the drainage system, they may, in the near future, create flash flooding during the raining season (Figure 1.2).

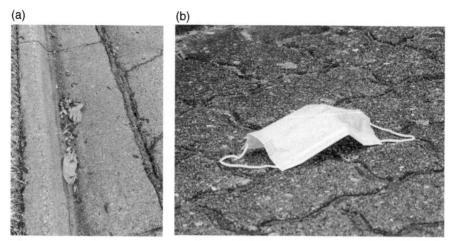

Figure 4.22 (a) Discarded protective gloves on the road. *Source:* Courtesy of SWIS. (b) Discarded protective mask on the street. *Source:* Photo by Matti Blume 2020. CC BY SA 2.0.

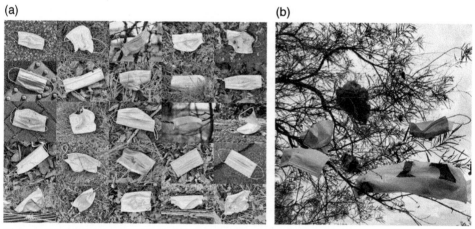

Figure 4.23 (a) Combinations of photos of Masks littering the environment. *Source:* Alan/Adobe Stock. (b) Coronovirus contributing to pollution, as discarded face masks clutter the environment along with other plastic trash. *Source:* Courtesy of SWIS.

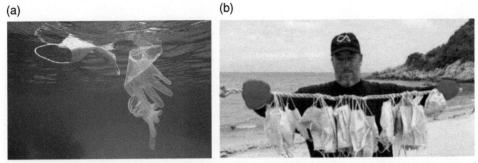

(a)

(b)

Figure 4.24 (a) Disposable masks end up in water (https://www.indiatvnews.com/news/india/coronavirus-disposable-face-masks-pose-serious-environment-threat-how-to-deal-with-it-669635). *Source*: damedias/Adobe Stock. (b) Littered facemasks near beaches (https://www.indiatimes.com/technology/science-and-future/covid-19-is-causing-rise-in-plastic-waste-that-will-harm-environment-warns-expert-511089.html). *Source*: Yoyo Chow / REUTERS/Adobe Stock.

4.3.3 Characteristics of Waste during COVID-19 (April to December 2020) (Aurpa 2021)

A study was conducted at the University of Texas at Arlington (UTA) to observe the pattern of changes in MSW generation due to the COVID-19 pandemic and its effect on landfills. Waste samples were collected from the Hunter Ferrell Landfill in Irving, Texas for seven months (April 2020 to December 2020). Eight, 9.0–13.6 kg (20–30 lb) bags of MSW samples were collected from the landfill to determine their physical properties: composition, moisture content, compacted unit weight, and volatile organic solid content. Changes in the waste's characteristics and an increase in the amount of plastic content were observed. Volume and weight characteristics of plastic waste during COVID 19 and its implication on life of landfill evaluated.

4.3.3.1 Characteristics of MSW

The percentage of each waste component by weight for all the months is provided in Figure 4.25. The percentage of the degradable paper, food waste, textiles, and wood and yard waste were approximately 60% by weight. The weight percentage of plastic more than doubled, from 13 to 30%. This was expected, based on the astronomical use of plastic materials during COVID-19 in 2020. The determined unit weight of solid waste (20 pcf) reflects this change in plastic content.

4.3.3.2 Plastic Waste Characterization

The plastic waste was further classified to identify the major uses of plastics during COVID 19. Figure 4.26 shows the classification of plastics both in weight and volume

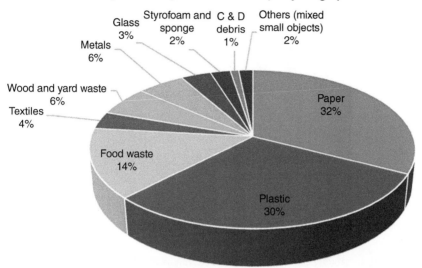

Physical composition of MSW (% by weight)

Figure 4.25 Percentage of different components of waste by weight in overall MSW.

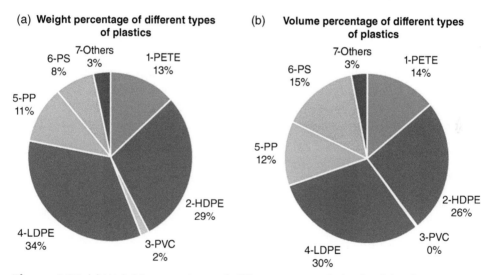

Figure 4.26 (a) Weight percentage of different types of plastics (b) Volume percentage of different types of plastics.

basis. LDPE (as packaging material) and HDPE (as milk and water bottles) were the main constituents of plastic waste found in the Hunter Ferrell Landfill.

Plastic wastes were compared with other wastes by considering plastic as a whole mass. The volume percentage of plastics with respect to the total volume of MSW was larger than

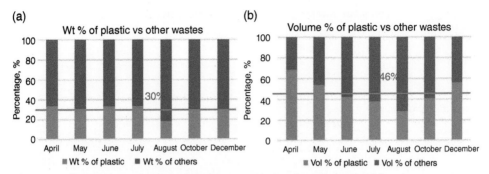

Figure 4.27 Monthly variation of (a) average plastic weight % with other waste percentage (b) average volume % with other waste percentage.

the weight percentage of plastics with respect to the total weight of plastics. Therefore, the volume of plastic waste takes up more space in landfills than its weight (Figure 4.27).

4.3.3.3 The Implication of Plastic Waste Increase on Landfill Life

An increase in plastic waste has a negative impact on the active life of a landfill. In other words, an increase in the volume of plastic waste reduces the active life of a landfill. The implications of the increased volume of plastic during the COVID-19 pandemic on the lifespan of landfills was studied by comparing the length of time that it took to fill up a landfill before and after the pandemic, and it was found that the pandemic reduced the lifetime of the landfill by 15%. The study only considered waste characteristics of one city in Texas and may not reflect the waste management practices of other U.S. cities or other countries; however, based on the collected data, the trend of higher plastic use and plastic waste generation was uniform across the globe (Figure 4.28).

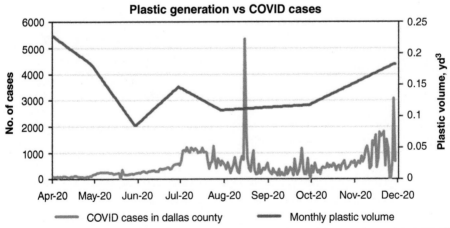

Figure 4.28 Comparison of plastic waste generation with number of COVID-19 cases in Dallas County during pandemic.

4.3.4 Reuse of Plastic Waste in Engineering Applications

The diversion of plastics from landfills also creates a market for recycled plastics (Hossain et al. 2017). Being nondegradable can be good if the plastics are reused for civil engineering infrastructure projects because they can last for a very long time. Therefore, developing sustainable waste management systems and optimizing the use of recycling materials require that local markets be created for recycling materials. This section focuses on plastic waste and its reuse, as it is viewed globally as a major polluter, environmental contaminant, and/or nuisance. Even before the COVID 19 pandemic, plastic production, consumption, and generation of plastic waste was widespread across the globe.

The use of plastic has increased significantly over the past few decades due to its relatively low production cost, high durability, and light weight. Plastic has a wide variety of applications, from short-term use (e.g. packaging, agricultural endeavors, and disposable consumer items) to long-term infrastructure materials (e.g. pipes, cable coatings and structural materials). Plastic can also be used for durable consumer applications with an intermediate lifespan, such as electronic goods, furniture, and vehicle parts. As a consequence of the ever-growing number of plastic products and the wide variety of their applications in modern daily life, the generation of post-consumer plastic waste has also been increasing rapidly (Plastics Europe 2013). Globally, approximately 1.3 billion tons (2.0 trillion lb) of waste are currently generated each year. By 2025, this number is projected to be 2.2 billion tons (4.9 trillion lb). An average of 10% of the main waste stream is plastic but varies according to the income level of the country. In low-income countries, the plastic content is found to be as low as 8%, whereas in high-income countries it comprises up to 11% of the total waste stream (Hoornweg and Bhada-Tata 2012).

Worldwide, the rate of plastic recycling is still minimal, with a maximum of 30% in Europe and only 10% in the U.S. (Plastics Europe 2013). The plastic that is not recycled/recovered is discarded into landfills. Unlike most other components of the total waste stream, plastic and plastic products do not degrade over time. In 2017, 35.4 million tons (78 billion lb) of plastic waste were generated in U.S., and only 8.5% was recycled; 15.8% was incinerated for energy recovery and 76% was sent to landfills (Figure 4.29).

The use of recycled plastics enhances sustainable resource management by reducing the environmental impact of using virgin material. Recycled plastic has extensive uses in many applications and products, some of which are described below (Hossain et al. 2017).

Construction: Recycled plastic has been widely used in mainstream construction products such as damp-proof membranes, drainage pipes, ducting, and flooring, due to its durability, low weight, low maintenance, resistance to vandalism, and non-degradability. Recycled plastic lumber is also used for pier construction. Another notable use of recycled waste plastic is the production of recycled plastic pins, which are used primarily to stabilize highway slopes.

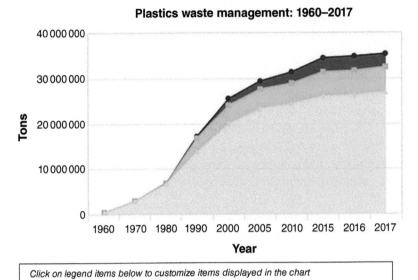

Plastics waste management: 1960–2017

Click on legend items below to customize items displayed in the chart
■ Recycled ▨ Composted ▨ Combustion with energy recovery ▨ Landfilled

Figure 4.29 Plastic waste management in the USA (USEPA)
Note: 1 ton = 2204.62 lb.

Packaging: Recycled PET and HDPE are widely used for packaging by retailers and brand manufacturers of plastic bottles and trays.

Apparel: Textile fibers for clothing, such as polyester fleece and polyester filling for duvets, coats, etc. are made from recycled PET bottles and is the largest market for recycled PET bottles worldwide.

Landscaping and street furniture: Recycled plastic has been used increasingly to build walkways, jetties, pontoons, bridges, fences, and signs. Traffic management products, such as street signs, street furniture, bins, and planters are also made from plastic.

Bags, sacks, and bin liners: New products, such as carrier bags, refuse sacks and bin liners, are made on a large scale from old plastic film from sources such as pallet wraps, carrier bags, and agricultural films.

4.3.4.1 Case Study I – The Use of Recycled Plastics Pins (RPPs) for Highway Slope Stabilization

The use of recycled plastics enhances sustainable resource management and completes the circular economy loop by reducing the environmental impact of using virgin material. Every year, 50 billion plastic water bottles end up in U.S. landfills, and it takes 500 years to decompose plastics. The use of RPPs made from recycled plastic bottles for highway slope stabilization and civil engineering infrastructure projects represents a perfect example of sustainable engineering solutions (Figure 4.30). RPPs are predominantly polymeric materials that are fabricated from recycled plastics and other waste materials and are composed of HDPE (55–70%), LDPE (5–10%), PS (2–10%), PP (2–7%), PET (1–5%) and varying amounts of additives, e.g. sawdust, fly ash (0–5%) (McLaren 1995; Lampo and Nosker 1997; Chen et al. 2007). RPPs are made of lightweight materials and are less susceptible to chemical and biological degradation than other structural materials. They require no maintenance and are resistant to moisture, corrosion, rot, and insects.

Other conventional civil engineering construction materials have the potential to degrade, corrode, or rot when they come in contact with moist soil or water (Figure 4.31).

When used for waterfront structures, jetties, or bridge foundations, wood degrades and loses its structural capacity or integrity, and eventually can cause failure. Columns made of recycled plastic, however, do not degrade or lose their structural integrity for a long period of time. Below-ground concrete or steel structures can corrode and lose their structural integrity when they come in contact in water, corrosive elements of soil, sulfates, or other minerals; however, recycled plastic pins do not degrade and their structural capacity is not affected by environmental pollutants.

Because plastic is not degradable, it is problematic when it is discarded into landfills. The same thing that makes it problematic for landfills, however, makes it useful and beneficial for civil engineering infrastructure projects, as soil and highway slopes repaired with RPPs retain their engineering characteristics for a long time, thereby reducing their overall maintenance and repair costs of slopes and pavement shoulders in the long run (Hossain et al. 2017). The following case study demonstrates the sustainable reuse of recycled plastics for civil engineering infrastructure projects.

A highway fill slope constructed during 2003–2004 is located on Highway US 287, near the St. Paul overpass in Midlothian, Texas (A photo of the slope is presented in

Figure 4.30 Plastic bottles can be recycled into recycled plastic pins (Hossain et al. 2017).
Source: Courtesy of SWIS.

(a)   (b)

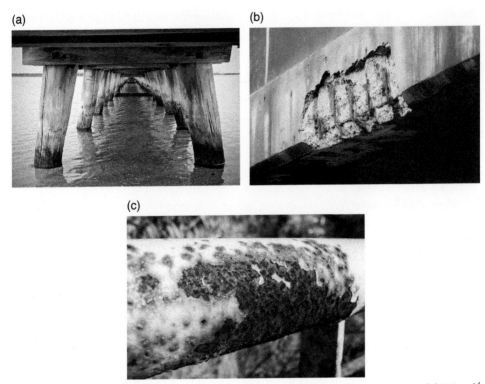

(c)

Figure 4.31 Conventional construction materials and their disadvantages (a) Wood/
Timber degrades (Substructure). *Source*: pxhere.com (b) Concrete-corrodes. *Source*:
Wirestock/Adobe Stock (c) Steel-corrodes. *Source*: Guilbaud Stan/Shutterstock.

Figure 4.32). The maximum slope height is about 9.14–10.7 m (30–35 ft), with a slope
geometry of 3 (H): 1(V). During September 2010, cracks were observed on the shoulder,
near the crest of the slope. This type of crack is known as a shrinkage crack and is
usually an early indication of slope movement and foretells slope failure within one
or two years, especially after a lengthy rainfall event. This slope was repaired using
recycled plastic pins.

Three 15.2 m (50 ft) wide sections over the US 287 slope, designated as Reinforced
Section 1, Reinforced Section 2, and Reinforced Section 3, were considered for
stabilization. Two unreinforced control sections that were between the reinforced
sections were also considered, to evaluate the performance of the reinforced section.
(The control slope means that it was not repaired). More details of the study can be
found at Khan (2014).

It was evident that the cracked zone was the initiation point of the critical slip
surface of the US 287 slope. Therefore, to resist the movement of the slope and
provide additional support, RPPs were installed close together at the crest of the

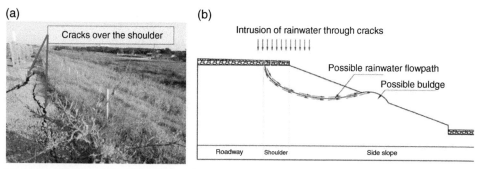

Figure 4.32 US 287 slope failure in Texas (a) Site photo (b) Schematic of water seepage from crest of slope.
Source: Khan 2014.

slope in Reinforced Section 1. More details of the study can be found at Khan (2014). Different pin layouts, lengths, and spacing were used for the slope repair to check their effect on the repair and the increase in the slope's stability (Figure 4.33).

A crawler-type drilling rig with a mast-mounted vibrator hammer (Klemm 802 drill rig, along with a KD 1011 percussion head drifter) was used during the current study to install the RPPs. The crawler-type rig is suitable for installation over slopes, as no additional anchorage is required to maintain the stability of the equipment, and it reduces the labor, cost, and time required for the installation process. The photographs of the RPP installations at the US 287 slope are presented in Figure 4.34. The RPPs were installed in April 2011 in Reinforced Section 1 and Reinforced Section 2.

The installations in Reinforced Section 3 were conducted in the following year (2012) and were spaced similar to those in Reinforced Section 2. Based on the study, the average driving rate, considering all three reinforced sections, was 0.81 m/min (2.66 ft/min), signifying that a 3.048 m (10 ft) long RPP can be installed within four minutes, or on average, 100–120 RPPs can be installed in one day. More detail about the equipment and the installation method can be found in Hossain et al. (2017).

To evaluate the performance of the reinforced slope, selected RPPs were instrumented with strain gauges. Three inclinometers were installed at Reinforced Section 1, the control section, and Reinforced Section 2 to monitor the horizontal movement of the slope (Figure 4.35). The depth of each inclinometer casing was 30 ft., and they were installed perpendicular to the slope surface, 6.096 m (20 ft) below the crest. The strain gauge and inclinometer showed the movement of the soil. After the RPPs were installed, the inclinometer data indicated that there was a substantial amount of movement in the control section, where no RPPs had been installed. Minimal movement appeared in the reinforced section, which indicated that RPP provided resistance and restricted the soil movement in the repaired area.

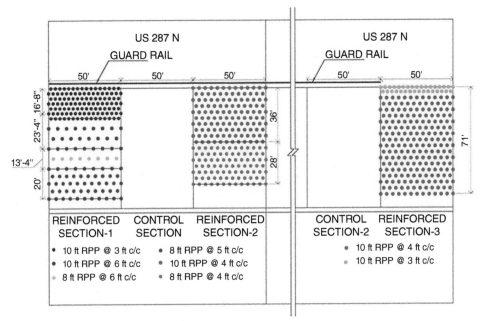

Figure 4.33 Layout of RPPs to stabilize the US 287 slope.
Source: Khan 2014.

(a)   (b)

Figure 4.34 Installation photos of RPP at (a) Reinforced Section 1 (b) Reinforced Section 2.
Source: Khan 2014.

A topographic survey was conducted over the US 287 slope to monitor the performance of the slope stabilization. The first survey was conducted after the RPP installation was completed in May 2012 and was continued on a monthly basis. During the survey, the cracked zones over the shoulder were monitored in different reinforced and control sections.

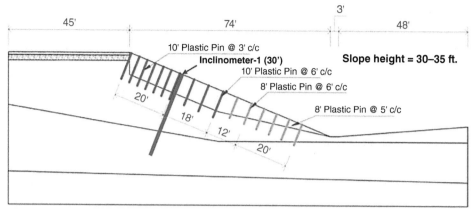

Figure 4.35 Cross section of Reinforced Section-1 at US 287 slope.
Source: Khan 2014.
Note: 1 ft. = 0.3048 m.

The total settlement over the crest of the slope as of 2021 is presented in Figure 4.36. The plot shows that the control sections of the slope had significantly more settlement at the crest compared to the reinforced sections. The maximum settlements were 15 in. (38.1 cm) and 9 in. (22.9 cm) in Control Section 1 and Control Section 2, respectively. This clearly indicates that the slope was moving in the control sections. On the other hand, there was very minimal settlement in the repaired area.

During the five years of the project, the northbound slope of highway US 287 was inspected visually on a monthly basis and a number of shallow slope failures were recorded on the control slope (northbound of US 287). The locations of the failures are presented in Figure 4.37. After the installation of the RPP, first-time failures (Location 1 and Location 2) on the northbound control slope were observed during September 2013, after a rainfall event. As this was a shallow failure with no significant damage to the landscape, no maintenance effort was needed. Later, in August 2014, the slope failed again in Location 3, after a heavy rainfall. Soil was backfilled within the failure location, which was a temporary solution to prevent the movement of the slope. A global failure (Location 4) of the slope was observed in June 2015 after a massive rainfall of 125 mm (4.92 in) during that month. As part of the slope repair, a soil nail wall was constructed to control the movement of the pavement, and the slope was reconstructed in October 2015 (Figure 4.38), when the highly plastic clay soil was stabilized with lime. Even after the reconstruction, the slope failed (Location 5) after a month of heavy rainfall, and after the repair, the slope failed again in the first week of November 2015 (Figure 4.39)).

During the failures of the northbound slope, a minor increment of horizontal displacement was observed at the inclinometer in Reinforced Section 1; however, no sign of failure was observed at the reinforced section.

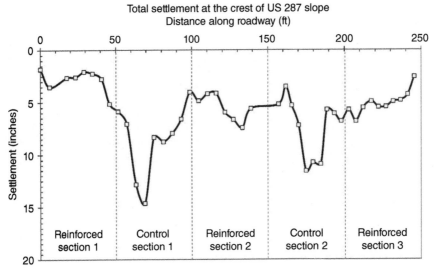

Total settlement at the crest of US 287 slope
Distance along roadway (ft)

Figure 4.36 Total settlements along the crest of US 287 slope.
Note: 1ft. = 0.3048m.

Failure location	Time
Failure location-1	October 2013
Failure location-2	October 2013
Failure location-3	June 2015
Failure location-4	June 2015

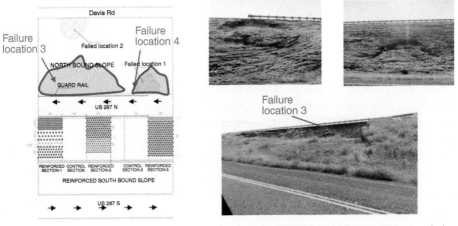

Figure 4.37 Schematic diagram of slope failures within northbound control slope (Hossain et al. 2017).
Source: Courtesy of SWIS.

(a)

September 2015

Soil nailed wall

(b)

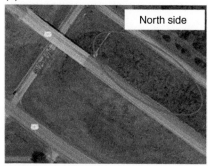
North side

October 2015

Figure 4.38 Slope repair work including soil nail wall (a) September 2015 (b) October 2015.
Source: Courtesy of SWIS.

Based on the monitoring results of slope stabilized with RPP, no failures occurred on any slopes repaired with RPPs whereas during the same period many slopes failed near the repaired sections. The repair with RPPs saved 70 to 80% of the cost of traditional methods and require a lot less time (one to two weeks using RPP versus three to four months for conventional slope repair methods). Based on the cost and environmental benefits, RPP provides tremendous potential for engineering applications and a sustainable resource management for recycled plastic reuse and an example of a circular economy for global practices.

The research results received national media attention after the successful repair of the slope on US 287 (Figures 4.40 and 4.41).

4.3.4.2 Case Study II: Plastic Road

Bitumen is a highly heterogeneous mixture of hydrocarbons and polymers that are similar to plastic, and asphalt roads can be constructed by mixing bitumen and aggregates. The plastic waste that is used for road construction is comprised of various items like bags; cups; and packaging for potato chips, biscuits, chocolates, etc. These plastic components of recycled plastic are very difficult to recycle and create a nuisance

Figure 4.39 Slope failure in first week of November 2015 (One month after the slope repair was completed) (Hossain et al. 2017).
Source: Courtesy of SWIS.

Figure 4.40 UTA researcher uses recycled plastic to strengthen highways (YouTube 2017).
Source: Solid Waste Institute for Sustainability (SWIS)., https://www.youtube.com/watch?v=c0DResP-F4I&t=6s.

in society by clogging drains, causing flooding, etc.; however, there are different types of plastic, and some of them are being investigated for road construction.

4.3.4.2.1 UTA Study on the Use of Recycled Plastic in Flexible Pavement Design

The rutting of flexible pavement that occurs at high temperatures is one of the main distresses that commonly occur in flexible pavement due to the permanent deformation of each layer of the pavement structure under repetitive traffic loading. Permanent deformation usually occurs due to the consolidation of either the underlying base course due to repeated traffic loads along the wheel path, or plastic flow near the pavement surface, which significantly reduces both the structural and functional performance of the pavement (Figure 4.42a). Fatigue cracking is another distress mechanism that causes degradation of pavements. It is caused by repeated traffic loadings, which result in crack initiation, crack propagation and eventually catastrophic failure of the material due to unstable crack growth (Figure 4.42b). The properties of a bituminous mix that can perform adequately under a wide range of temperatures and offer resistance against degradation caused by stresses and loads must be compatible to offer thermal stability; load spreading; and chemical stability to prevent the pavement from rutting, fatigue, and thermal cracking. Modification of bitumen with a synthetic polymer binder can provide a new approach to overcoming new technical demands.

Workers insert plastic pins underneath Texas Route 287 as part of a pilot program that concluded this summer (Courtesy Sahadat Hossain/UTA/TxDOT)

Figure 4.41 What if roads lasted twice as long? (Grabar 2013; https://www.citylab.com/transportation/2013/09/what-if-roads-lasted-twice-long/6862).
Source: Courtesy Sahadat Hossain/UTA/TxDOT.

Increased traffic loads and high pressure resulting from heavy vehicles have been among the factors causing rutting, cracking and premature failure of pavements. These must be given careful consideration in the design of the pavement and selection of materials, to preclude premature failure of bituminous pavements. The lack of stability in an asphalt mixture also causes the road's surface to unravel and rut (Ahmad 2014).

Plastics account for a considerable amount of the solid waste in the world due to the many ways that are not reused or recycled from sources such as in packaging, building and construction, automotive, electric, and electronic applications (Gawande et al. 2012). Since they have a high decomposition temperature, high resistance to ultraviolet radiation, and are mostly not biodegradable, they can remain on both land and sea for years, causing environmental pollution. As plastic usage is increasing day by day due to population growth, urbanization, development activities, and changes in lifestyle, it is difficult to dispose of (Venkat 2017) and is either landfilled or incinerated, neither of which is eco-friendly and both of which pollute the land and air (Shiva et al. 2012).

The concept behind this research was to address these two major problems simultaneously. If recycled plastic could be used as bitumen modifier it would be a sustainable and cost-effective solution for fixing distressed pavement as well as reducing plastic pollution. The objective of this study was to evaluate the potential reuse of recycled plastics as a partial replacement of bitumen in flexible pavement design.

4.3.4.2.2 Why Plastic Roads?

When bitumen is added to an aggregate that is mixed with plastic, better adhesion is formed between the bitumen and the plastic-coated aggregate due to a strong intermolecular bonding. The visco-elastic property of the mix does not change over time due to the strong intermolecular attraction between bitumen and plastic-coated aggregate that enhances the durability and stability the of pavement. Hence plastic roads do not show any evidence of change in their visco-elastic nature, even with prolonged exposure to the atmosphere and under different environmental conditions. This helps in increasing the stability of the mix and reducing the stripping of bitumen that results in the surface layers unraveling and loosening. Plastic-bitumen composite roads have better wear resistance than standard asphalt concrete roads. They do not absorb water and are more flexible, which results in less rutting and fewer needed repairs. They are unaffected by corrosion and weather and can handle temperatures from −40°F to 176°F with no negative effects. In summary, plastic-waste-modified-bitumen asphalt mixes are expected to be more durable, less susceptible to moisture and extreme temperatures, and exhibit improved performance in actual field conditions (Sabina et al. 2009).

4.3.4.2.3 Types Of Plastics

There are seven grades of plastics, the most common of which are polyethylene terephthalate (PET), polyethylene (PE), polyvinyl chloride (PVC), polypropylene (PP),

Table 4.2 Plastic grades, their common uses, and properties.

Plastic grade	Plastic type	Common uses	Properties
1	Polyethylene terephthalate (PET)	Plastic bottles (water, soda drinks, etc.)	Clear, strong, and lightweight
2	High-density polyethylene (HDPE)	Milk containers, shampoo bottles, cleaning agents, etc.	Stiff and hardwearing, hard to breakdown in sunlight
3	Polyvinyl chloride (PVC)	Plastic piping, vinyl flooring, cabling insulation, roof sheeting	Can be rigid or soft via the use of plasticizers; used in construction, healthcare, electronics
4	Low-density polyethylene (LDPE)	Plastic bags, plastic food wrapping (e.g. fruits, vegetables)	Lightweight, low-cost, versatile; fails under mechanical and thermal stress
5	Polypropylene (PP)	Bottle lids, food tubs, furniture, automobile parts, etc.	Tough and resistant, effective barrier against water and chemicals
6	Polystyrene (PS)	Plastic cutlery, containers for "To Go" orders	Lightweight, structurally weak, and easily dispersed
7	Others	Fiberglass, water cooler bottles	Diverse in nature with various properties

Source: Based on Tsakona and Rucevska 2020.

and polystyrene (PS), which represent 69% of all plastics globally (Geyer et al. 2017). Table 4.2 depicts the various plastic grades, their common uses, and properties.

4.3.4.2.4 *Plastic Generation and Recycling*

According to the Environmental Protection Agency (EPA), 35.7 million tons of plastics were generated in the United States in 2018. That was 12.2% of all of the MSW generated. The United States recycled 3.1 million tons of the plastic, which represents only 8.7%. The recycling rate of plastic is sinking daily since China banned the importation of foreign waste from many countries, including the U.S. in January 2018. Since then, plastics that were previously recycled are now going to landfills, where they occupy a large volume of space for a long time.

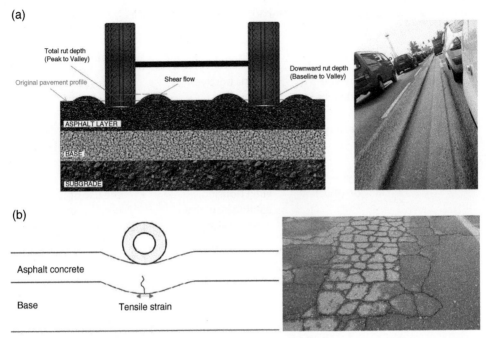

Figure 4.42 (a) Rutting. *Source*: Courtesy of SWIS; Burda/Wikipedia Commons/ Public Domain. (b) Fatigue cracking (*Source*: Islam, 2015).

While the number of recycled plastics overall is relatively small, recycling of some specific types of plastic containers such as PET, HDPE, PP is more significant. There are few, if any, recycling facilities for low-density polyethylene (LDPE) such as grocery bags, plastic wrap, and Polystyrene (PS). Figure 4.43 shows the types of recyclable and non-recyclable plastics.

4.3.4.2.5 Plastic Types Suitable for Plastic Road

Using the wrong type of plastic can have adverse results, such as premature failure of pavement (Hınıslıoğlu and Ağar 2004; Montanelli 2013; Delongui et al. 2018). Literature shows that polymer materials that can be used are low-density polyethylene, such as plastic bags, films, foams, high-density polyethylene, and polypropylene. It is difficult to mix PET and polystyrene particles with bitumen because of their high melting point (usually greater than 200° C) (Modarres and Hamedi 2014). Vasudevan et al. (2010) stated that polyvinyl chloride (PVC) should not be used in order to prevent the possibility of chlorine contamination. To consider all of these things, the current research aims to use LDPE, HDPE, and PP for plastic road construction. One-time-use plastic bags are considered non-recyclable plastic, as they are difficult to recycle and create problems for the environment and society. Since LDPE (e.g. plastic grocery bags, plastic wraps,

Figure 4.43 Plastic recycling codes infographic.
Source: elenabsl/Adobe Stock.

plastic films) does not have any recycling value and ends up in landfills, it would be a good source for plastic pavements. HDPE and PP can also be used in pavements, as they can be easily melted under 200° C without any adverse effects on the environment.

4.3.4.2.6 Issues That Need to Be Addressed

Recycled plastic has been used in flexible pavements in many countries around the world to increase the stability and durability of roads and to reduce the cost of construction by replacing a percentage of the bitumen with waste plastic. The lack of widely accepted mix design limits its use globally; however, the following need to be pursued:

- Investigation of type of plastics that can be used to modify the bitumen.
- Evaluation of different characteristics of plastic-modified bitumen and its effect on short-term and long-term performance of flexible pavement
- Evaluation of environmental effects of using plastics in bituminous roads.

The current research will address the above-mentioned issues and identify the correct mix design for plastic roads. The preliminary test results revealed that rutting can be significantly reduced by adding plastics to roads (as presented in Figure 4.44).

Engineers must understand the behavior of plastic-modified asphalt pavement if they are to produce sustainable and economical designs and construction procedures and governmental agencies must be willing to modify their road specifications to

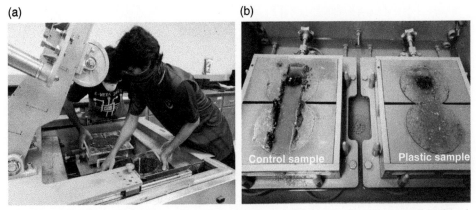

Figure 4.44 (a) Sample placement for Hamburg Wheel Tracker Testing (b) Visual observation of rutting for both control sample (without plastic) and plastic sample using Hamburg Test.
Source: Courtesy of SWIS.

incorporate these designs. Therefore, evaluating the effect of different plastic types (e.g. LDPE, HDPE, PP etc.) is important for their effective application in reducing pavement distresses.

4.3.5 Reuse of Recycled Food Waste: Composting

More than 50% of the MSW in a developing country is market waste, food waste, kitchen waste and other organic waste that is suitable for composting. Composting is a major component of sustainable waste/resource management; however, organic waste components of MSW are often mixed with other inorganic materials and recyclables, glass, plastics, metals, and household hazardous materials which can potentially contaminate the final composting product if they are not separated from organic waste before being composted. Separating the food waste/kitchen waste/vegetable market waste at its source is a major challenge, but without proper separation, the compost is likely to be contaminated. The separation of contaminants can be accomplished in two different ways: (i) Source separation at the household level. and (ii) separation of the organic waste in MRFs, as described in previous sections. It is within the power of city officials and regulators to mandate that market waste (which is mainly organic waste) be collected separately from household waste and directly used for composting purposes, minimizing the contamination level as much as possible.

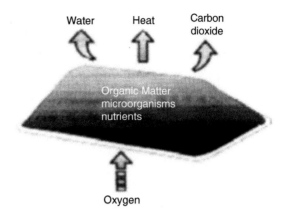

Figure 4.45 Composting inputs and outputs (Hoomweg et al. 1999).

In natural environments, the slow decomposition of organic matter by different microorganisms produces humus, a black-brown earthy material that is a valuable component of good soils. Composting is the natural biological degradation of organic matter in a more systematic and scientific way to enhance biological degradation. Microorganisms convert the organic matter into a humus-like material, and the end product is compost (Hoomweg et al. 1999). The following general formula and Figure 4.45 illustrate the inputs and outputs for the conversion of organic matter in the presence of oxygen:

$$\text{organic matter} + O_2 + \text{nutrients} \xrightarrow{\text{bacteria}} \text{new cells} + \text{resistant organic matter} + CO_2 + H_2O + NH_3 + SO_4^{2-} + \ldots + \text{heat}$$

(Tchobanoglous et al. 1993)

Composting is a relatively simple process that is an important part of diverting waste from final disposal. Organic components of MSW waste are readily decomposable, although some materials are more suitable for composting than others. The raw materials that are most appropriate include vegetable and fruit waste; farm waste such as coconut husks and sugar cane; crop residues such as banana skins, and corn stalks and husks; yard waste such as leaves, grass, and trimmings; sawdust; bark; household kitchen waste; human excreta and animal manure (Hoomweg et al. 1999). Wood, bones, green coconut shells, paper and leather decompose very slowly and hinder the composting process, ideally, they should be removed from the composting pile (Lardinois and van de Klundert 1993).

4.3.5.1.1 Case Study: Community-Based Composting in Bangladesh

4.3.5.1.1.1 Introduction

In the rapidly growing city centers of Bangladesh, waste management has become extremely problematic. Existing landfills are almost at capacity, and developing new landfills remains a challenge due to the scarcity and cost of land. Although local governments spend about 5 to 20% of their annual budget on waste management, 35 to 50% of the total waste generated remains uncollected. The waste sector is a significant contributor of greenhouse gas emissions in Bangladesh, and in 2005, Bangladesh generated about 4.86 million tons (10.7 billion lb) of waste. Dhaka, the capital city of Bangladesh, generated about 4700 metric tons (10 million lb) of municipal solid waste per day in 2010 (Rahman 2011). About 80% of waste generated in Dhaka is organic, which provides a major opportunity for recycling it into compost (Enayetullah and Hashmi 2006), which is in high demand for use as an organic fertilizer for agriculture. Furthermore, composting reduces the amount of methane generated from organic waste by diverting it away from landfills.

4.3.5.1.1.2 Community Actions

Waste Concern, a social business enterprise, initiated the idea of converting food waste into resources in Dhaka. They developed a network of decentralized composting plants that were adapted to the local Bangladeshi context and were financially self-sufficient. This initiative created a solution that has the potential to sustainably manage municipal solid waste; reduce costs to the city; reduce methane emissions from waste; and create employment opportunities for the urban poor, who have traditionally earned a living by sorting through trash to collect and sell recyclables.

In developing countries, large, centralized composting plants are often not economical due to their high operational, maintenance, and transportation costs, and cities often lack the resources to administer such services. Small, decentralized composting locations can be a practical alternative because they require fewer technological and financial resources, albeit a greater number of laborers.

Waste Concern launched its first pilot project in 1995 in Dhaka. It provided communities with door-to-door waste collection for a monthly fee that was based on affordability. The pilot project's success persuaded the government of Bangladesh to partner with them to replicate the model on four sites through the United Nations Development Program (UNDP). However, Waste Concern had to overcome a critical barrier - land acquisition. In line with two government policies issued in 1998, Dhaka provided public land for siting compost units, as well as water and electricity connections, to composting facilities at no cost. Waste Concern was responsible for training the community how to manage, operate, and maintain the composting facilities, as well as market the product. It also monitored the composting sites for an additional three years to ensure sustainability and provided technical support when it was needed.

The composting plants produced quality compost fertilizer from organic waste collected from households and vegetable markets. Waste was taken to a sorting center to separate out the recyclables and to process the organic waste into compost (Figure 4.46). It takes less than 60 days to convert organic waste into compost using a labor intensive, aerobic, low-cost technology, and the sale of recyclables and compost generates enough revenue to pay the workers and maintain and operate the units. It is important to note that one ton of organic waste produces approximately 15–20% of end-product compost and reduces a half-ton of methane emissions.

4.3.5.1.2 Benefits of Composting

Some of the major benefits of composting as presented in World Bank white paper (1999) are listed below:

- It maximizes waste diversion from landfilling, as more than 80% of MSW in developing economies can be organic.
- If the organic waste can be diverted from the waste stream, it increases the potential for recycling and incineration operations.

Figure 4.46 The maturing shed at the Bulta Composting Facility.
Source: CENTER FOR CLEAN AIR POLICY., https://ccap.org/assets/CCAP-Booklet_BangladeshCompost.pdf.

- It produces a valuable soil amendment that is integral to sustainable agriculture.
- It promotes environmentally sound practices, such as the reduction of methane generation at landfills.
- It enhances the effectiveness of fertilizer applications.
- Depending on the organic waste separation and transportation policies in place, it can potentially reduce waste transportation requirements.
- It requires very little start-up capital or operating funds, which can encourage young entrepreneurs to get into the composting business.
- It addresses significant health effects that result from organic waste, such as reducing dengue fever.
- It provides an excellent opportunity to improve a city's overall waste collection program by integrating the informal section and collection and source separation and using the organic waste for composting.
- It can create hundreds of jobs in every city and make use of the talents of young business professionals by introducing them to waste management.

It needs to be noted that the major weak spot in composting in almost every developing economy is the source separation of the organic waste from other waste components.

4.4 PART C: SUSTAINABLE WASTE MANAGEMENT FRAMEWORK – DISPOSAL/ FINAL DESTINATION

4.4.1 Anaerobic Digester

The food waste that is diverted from MRFs by source separation or collected market organic waste can be utilized for energy, such as heat generation through community-level or household-level anaerobic digesters (AD). In developing economies, more than 50% of the waste produced is organic and biodegradable, so AD can be a sustainable and ecofriendly option for organic waste management. Many countries are trying to reduce their over-reliance on fossil fuels because of both environmental and supply concerns, and are looking for alternatives to wood as fuel, the main energy source for cooking and heating as this is a major cause of severe deforestation. Wood, dried cow dung, dried tree branches, and dried rice husks (commonly used for cooking in many countries) are major sources of indoor air pollution due to the smoke that is emitted during cooking when they are used as fuels.

Approximately half the world's population and up to 90% of rural households in developing countries are still dependent on natural biomass fuels in the form of wood, dung, and crop residues (Bruce et al. 2000) which are typically burned in ineffectively functioning stoves or open fires indoors. This combustion process is fragmented in most of the cases, which results in significant emissions in the presence of poor ventilation and produces very high levels of indoor pollution (WHO 2000)

   i. In developing countries, the mean 24-hour levels of carbon monoxide are in the range of 2–50 ppm in homes using biomass fuels. The United States' Environmental Protection Agency's eight-hour average carbon monoxide standard is 9 ppm, but values of 10–500 ppm have been reported during cooking (USEPA 1998).

   ii. Indoor concentrations of particulate matters usually exceed guideline levels by a large margin. The 24-hour levels mean that the PM_{10} levels are typically in the range 300–3000 mg/m^3 and may reach 30000 mg/m^3 or higher during periods of cooking (Smith et al. 1994). The United States EPA standard for 24-hour average PM_{10} concentrations is 150 mg/m^3 (USEPA 1998).

   iii. Women's exposure to emissions is much higher than men's, as they do most of the cooking in developing countries (Behera et al. 1988) (Figure 4.47). The effects of the emissions begin with runny eyes and noses and sore throats, and gradually start to affect the respiratory system. Very common syndromes to respiratory illnesses are asthma, dyspnea, and palpitations (USEPA 1998). The exposure to benzopyrene that is emitted from stoves that are used for cooking an average of three hours every day can be compared to smoking two packets of cigarettes daily (Bruce et al. 2000). Moreover, mothers who carry young children on their back while cooking expose them to the benzopyrene (Albalak 1997). Additionally, research shows that children breathe in more air in proportion to their weight than adults and as young children spend proportionally more time indoors, that results in their developing various respiratory illnesses (Albalak 1997). Thus, for example, while the number of alveoli at birth is approximately 24 million, by the age of four, this number increases to 250 million (Bruce et al. 2000). Additionally, acute lower respiratory infections are found to be the single most important cause of mortality in children aged under five years, accounting for around 2 million deaths annually in this age group (Bruce et al. 2000). The indoor cooking and associated pollutants are major causes of

Figure 4.47 Indoor cooking in rural households using natural biomass fuels in (a) Nepal (https://www.washingtonpost.com/opinions/these-cheap-clean-stoves-were-supposed-to-save-millions-of-lives-what-happened/2015/10/29/c0b98f38-77fa- 11e5-a958-d889faf561dc_story.html). *Source*: Karan Singh Rathore/Flickr. (b) Africa (https://www.renewablesinafrica.com/biofuels-a-versatile-power-option-to-accelerate-energy-transition-in-post-covid-africa). *Source*: davide bonaldo/Adobe Stock. (c) Typical wood fired stove in developing countries (www.downtoearth. org.in/blog/pollution/air-pollution-in-rural-india-ignored-but-not-absent-75341). *Source*: Wikimedia Commons/Public domain.

respiratory diseases in both rural areas and developing countries. Providing alternative sources of cooking fuels or tools can significantly reduce the indoor air pollution and can save the lives of millions of children and women around the world in poor nations.

Biogas, generated from anaerobic digestion, using organic waste, provides a unique alternative for cooking and aids in sustainable organic waste management. There cannot be a better example of sustainability than when organic waste, which is viewed as a potential liability in many countries, is converted into a sustainable solution and an alternative to wood as potential fuel. Cooking, using biogas from an AD, does not cause smoke and helps improve indoor air quality. Anaerobic digestion of organic waste provides many benefits by: (i) improving air quality for rural and poor communities through the use of biogas for cooking, (ii) saving lives of women and children, (iii) generating renewable energy, (iv) reducing greenhouse gases, (v) reducing dependency on fossil fuels, (vi) reducing solid waste volumes and thus waste disposal costs, (vii) contributing to the preservation of natural resources by reducing deforestation, (viii) creating jobs, and (ix) closing the nutrient cycle. The added revenue streams and financial savings from the production of biogas and commodities also create social, health, and financial benefits for communities (GMI 2013).

Anaerobic digestion is a microbiological process that uses bacteria to break down organic matter in the absence of oxygen. An engineered system is used to design and operate an AD system (Figure 4.48) that contains an airtight chamber that serves as a

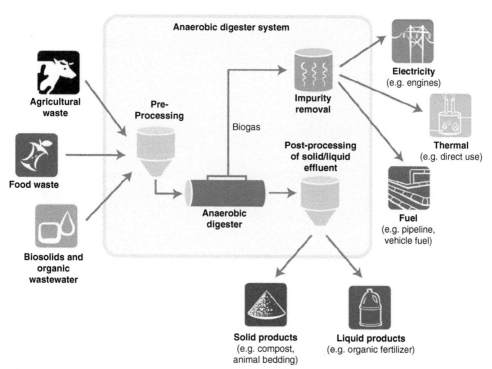

Figure 4.48 Anaerobic digestion.
Source: Modified from GMI 2013.

digester, where manure, biosolids, food waste, other organic wastewater streams, or combinations of these feedstocks decompose and produce biogas, a blend of methane and carbon dioxide, and digestate (GMI 2013).

Calorific value of biogas – The calorific value of biogas can vary depending on the raw materials used for biogas generation and is usually around 6.0–6.5 kWh/m³. (Deublein and Steinhauser 2011). The net calorific value depends on the efficiency of the biogas burners or other appliances used to process the biogas.

Table 4.3 shows examples of calorific values of different fuel sources as compared to biogas, as well as the approximate mass of each fuel, corresponding to 1 m³ of biogas.

The rule of thumb is that roughly 10 kg (22.0 lb) (wet weight) of biowaste (e.g. kitchen and market waste) are needed to produce 1 m³ (35.3 ft³) of biogas. This amount of biogas contains approximately 6 kWh (or 21.6 MJ) of energy.

Feedstock is mostly organic waste. Historically, anaerobic digestion has been used to treat liquid wastes with low total solids (less than 15%), such as manure, sewage, industrial wastewater, and sludge from biological or physical–chemical treatment. Today, however, organic components of solid wastes, such as agricultural and municipal solid waste, are mixed with different types of sludge or enzymes which are used as feedstock in the anaerobic digestion for high potential for biogas production (EAWAG). The presence of volatile solids (VS) in the total solids (TS) of food waste contributes to the biogas production (Table 4.4).

The amount of biogas produced depends on many factors, including type and composition of feedstock, type of solid waste organic, temperature, operational method, skill level and/or training of the operator, and different combinations. The biological methane potential (BMP) is the most common indicator of AD performance. BMP estimates

Table 4.3 Calorific values of different fuel sources (EAWAG).

Fuel source	Approximate calorific value	Equivalent to 1 m³ biogas (approx. 6 kWh/m³)
Biogas	6–6.5 kWh/m³	
Diesel, Kerosene	12 kWh/kg	0.50 kg
Wood	4.5 kWh/kg	1.30 kg
Cow Dung	5 kWh/kg dry matter	1.20 kg
Plant residues	4.5 kWh/kg dry matter	1.30 kg
Hard coal	8.5 kWh/kg	0.70 kg
Propane	25 kWh/m³	0.24 m³
Natural gas	10.6 kWh/m³	0.60 m³
Liquefied petroleum gas	26.1 kWh/m³	0.20 m³

Note: 1 kg = 2.2 lb.

Table 4.4 Total Solids (TS) and Volatile Solids (VS) in biowaste.

Substrate	TS (% of raw waste)	VS (% of TS)	Literature source
Spent fruits	25–45	90–95	Deublein and Steinhauser (2011)
Vegetable wastes	5–20	76–90	Deublein and Steinhauser (2011)
Market wastes	8–20	75–90	Deublein and Steinhauser (2011)
Leftovers (Canteen)	9–37	75–98	Deublein and Steinhauser (2011)
Overstored food	14–18	81–97	Deublein and Steinhauser (2011)
Fruit wastes	15–20	75–85	Gunaseelan (2004)
Biowaste	25–40	50–70	Eder and Schulz (2007)
Kitchen waste	9–37	50–70	Eder and Schulz (2007)
Market waste	28–45	50–80	Eder and Schulz (2007)

Table 4.5 Methane yield of organic component of solid waste.

Substrate	Methane Yield (L/kg VS)
Palm oil mill waste	610
Municipal solid waste	360–530
Fruit and vegetable wastes	420
Food waste	396
Rice straw	350
Household waste	350
Swine manure	337
Maize silage and straw	312
Food waste leachate	294
Lignin-rich organic waste	200

Note: 1 L/kg = 0.119 gal/lb.

the maximum possible volume of methane gas that can be produced per unit mass of solid or volatile solid matter (Buffière et al. 2006). Some methane yield values from anaerobic digestion of solid organic waste are shown in Table 4.5. The average methane yield of MSW is between 0.36 and 0.53 m^3/kg (5.8–8.5 ft^3/lb) VS (Khalid et al. 2011).

Biogas can be utilized for many different applications. In developing countries, it is most commonly used for cooking or engines. The amount of biogas needed to cook for one hour with a simple household stove in Tanzania and the corresponding consumption rate for other household burners are given in Table 4.6.

Table 4.6 Biogas consumption rate.

Biogas Application	Consumption rate (l/h)
Household cooking stove	200–450
Industrial burners	1000–3000
Refrigerator (100 l) depending on outside temperature	30–75
Gas lamp, equivalent to 60 W bulb	120–150
Biogas/Diesel engine per brake horsepower (746 W)	420
Generation of 1 KWh of Electricity with biogas/Diesel mixture	700

Note: 1 L = 0.26 gal.

Table 4.7 Amount of gas used to cook different dishes in India.

Item	Gas required (liters)	Time (Min)
1 l water	30	10
3 l water	75	25
500 g rice (3 l water)	205	65
500 g rice (3 l water) with thermal cooker	105	35
10 rice pancakes ("Appam")	70	25
Two steamed rice cakes	45	15

Note: 1 l = 0.26 gal.

The daily gas demands of families vary according to their diet and social habits. Research in Nepal revealed that 400 l of biogas is typically consumed per hour by locally constructed household stoves (Lohri et al. 2010).

Table 4.7 gives some indication of how much gas is used to cook different dishes in India. The gas flow rate of the stove was about 180 l/h.

Based on research conducted at UTA, organic waste/food waste can be mixed with sludge or other nutrients in various ratios to optimize the gas production.

4.4.1.1 UTA Research on Gas Production (Latif 2021)

A research study was conducted at UTA with the specific objective of determining the best combination for household-level AD in developing countries, using food waste. Most grains are produced in South Asia, Southeast Asia, and North Africa, as presented in Figure 4.49. Fruits and vegetables are also produced in significant amounts, followed by dairy products, meat, and fish. Considering this production scenario, two combinations of food waste were selected for the current study: grain and meat, and fruit and vegetables.

An experimental program was designed, and laboratory scale reactors were simulated as Ads (Figure 4.50). The degradation process, gas production, leachate, and

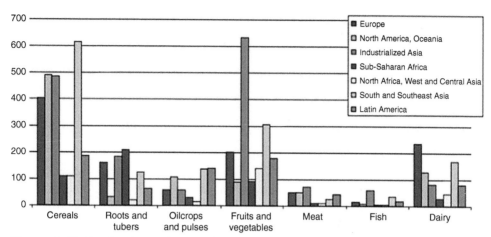

Figure 4.49 Production volumes of each commodity group per region (million tons) (FAO 2011; Gustafsson et al. 2013).

Figure 4.50 Schematic diagram for reactor setup.

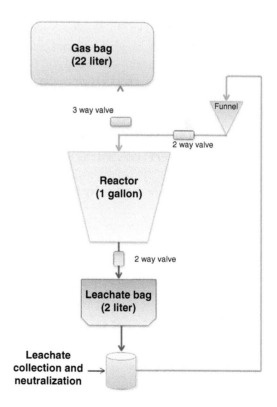

Table 4.8 Waste and inoculum combination for reactor feedstock.

Reactors	Food waste	Sludge
MGR1, MGR2	40% meat + 40% grains	20%
FVR3, FVR4	40% fruits + 40% vegetables	20%
MGR5, MGR6	35% meat + 35% grains	30%
FVR7, FVR8	35% fruits + 35% vegetables	30%

other parameters were monitored for 200 days. The composition of food waste and inoculum for feedstock of each reactor is presented in Table 4.8.

4.4.1.1.1 Sample Collection and Storage

Food waste was collected from two sources:

1. Fruits and vegetables were collected from the UTA Compost Center.
2. The meat and grains were collected from the lunch buffet in the University Center dining hall (Connection Café) at the University of Texas at Arlington (UTA).

Approximately 9.1 kg (20 lb) of food waste comprised of meat and grain products (rice, noodles, pasta, bread, etc.) were collected from the café. The collected samples were brought to the laboratory and stored at 4 °C (38 °F) in the environmental growth chamber to preserve their original properties before building the reactors (Figure 4.51).

4.4.1.1.2 Collection of Sludge

A 20 L (5-gal) bucket was used to collect the sludge from the Village Creek Water Reclamation Facility in Arlington, Texas, as shown in Figure 4.52, and the sludge was added to the reactors as the micro-organism source.

4.4.1.1.3 Production of Simulated AD Gas

During the monitoring period, the highest amount of gas was generated in reactors FVR3 and FVR4 (25.3 l/lb) that contained fruits, vegetables and 20% sludge, followed by reactors FVR7 and FVR8 (18 l/lb) containing fruits, vegetables, and 30% sludge. The total volume of gas generation from all of the reactors is shown in Figure 4.53.

The cumulative methane production data is presented in Figure 4.54. A similar trend was observed on the cumulative gas and time graphs, except the methane generation did not begin until after the lag phase. A considerable amount of methane was observed in the reactors with fruit and vegetables, with lag period between 70 and 100 days, 70 days for 30% sludge addition, and 100 days for 20% sludge addition. The volume of

(a)   (b)

(c)   (d)   (e)

Figure 4.51 (a) Collection of fruit and vegetable waste. *Source*: Courtesy of SWIS. (b) Collected of fruit and vegetable waste. *Source*: Courtesy of SWIS. (c) Collected meat and grain waste from UTA. *Source*: Courtesy of SWIS. (d) Storage of waste samples. *Source*: Courtesy of SWIS. (e) Environmental control chamber. *Source*: Courtesy of SWIS.

methane generated in the fruit and vegetable reactors varied between 3.5 and 7.5 l/lb, whereas the methane production was insignificant in the reactors with meat and grains. A considerable amount of VFA (volatile fatty acid) accumulation was observed in the meat and grain reactors, causing minimal production of methane in those reactors.

4.4.1.1.4 Gas Yield

The rates of total gas and methane generation are shown graphically in Figures 4.55 and 4.56. The peak value of the methane yield observed from reactors FVR3 and FVR4 was 343 ml/lb/day and 307 ml/lb/day, respectively, followed by reactors FVR7 and FVR8 for which the methane yield was 181.7 and 162.9 ml/lb/day, respectively.

(a) **(b)**

Figure 4.52 (a) Collection of sludge. *Source:* Courtesy of SWIS. (b) Collected sludge. *Source:* Courtesy of SWIS.

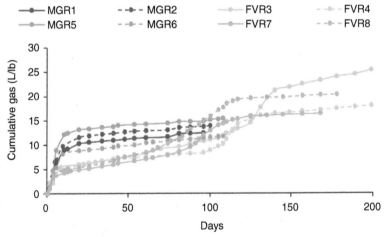

Figure 4.53 Cumulative gas generated versus time for all reactors. *Note:* 1 l/lb = 2.2 l/kg.

4.4.1.1.5 Summary

Based on this limited data, it can be concluded that combinations of fruits, vegetables, and sludge should be used for community AD operations and addition of meat and grains should be avoided.

4.4.2 Temporary Disposal (Biocell)

Based on the sustainable waste management flow chart, during MSW processing the waste left over after recycling, composting, and AD goes for temporary disposal in

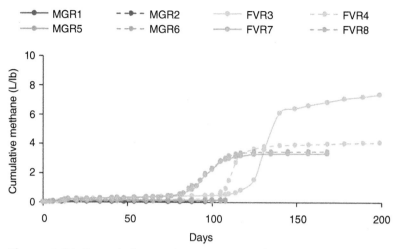

Figure 4.54 Cumulative methane generated versus time for all reactors.
Note: 1 l/lb = 2.2 l/kg.

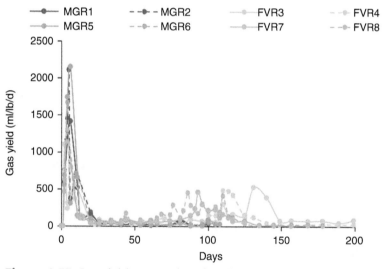

Figure 4.55 Gas yield versus time for all reactors.

(i) A perpetual landfill (in the form of a biocell), and/or (ii) a waste-to-energy plant. The type of disposal depends on the type of waste characteristics of a specific location. This section focuses on temporary disposal in perpetual landfills in developing countries, and another section will focus on the feasibility of WTE in developing countries and how each can work.

The concept of a perpetual landfill has been explained in previous sections. A perpetual/sustainable landfill (biocell) with a complete up-front removal of plastics, glass, and metal, can be used in one location in perpetuity to generate renewable

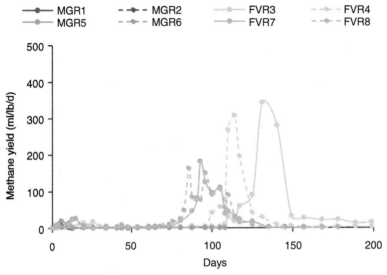

Figure 4.56 Methane yield versus time for all reactors.
Note: 1 ml/lb/day = 2.2 ml/kg/day.

energy, as organic waste completely degrades. Since non-degradable plastics, glass, and metals are removed up front, all of the waste placed in the biocell can be degraded. Therefore, any materials that remain after degradation are organic in nature and can be reprocessed for use as compost, or at the landfill as an alternative daily cover material or as backfill soil during construction. This section will focus on different types of landfilling operations and how biocell or perpetual landfilling can be implemented. UTA's laboratory and field experimental research findings on biocells will also be presented.

In conventional landfills that are designed and operated in accordance with the USEPA Subtitle D Regulations of the Resource Conservation and Recovery Act (RCRA), efforts are typically made to minimize the amount of moisture that enters the landfill. This is to minimize the generation of leachate and reduce the risk of groundwater contamination. However, the time required for the decomposition of waste in a dry tomb landfill ranges typically from 30 to 100 years, and the landfill gas is produced at a slow rate over a long period of time. Traditional landfilling, an improvement on random open dumping, is not a long-term sustainable solution and has negative impacts on the environment and urban sustainability.

In the mid-1970s, Pohland (1975) proposed the idea of enhancing waste decomposition by recirculating the leachate and/or the addition of supplemental water. Additional moisture stimulates microbial activity by providing better contact between insoluble substrates, soluble nutrients, and microorganisms (Barlaz et al. 1990). As a result, decomposition, and biological stabilization of MSW can be reduced to years rather than decades for traditional dry landfills. A major aspect of a bioreactor landfill operation is the recirculation of collected leachate back through the refuse mass. The concept of a bioreactor landfill operation is just opposite that of a traditional landfill operation in terms of moisture intrusion into the landfill. A bioreactor landfill is operated to enhance refuse decomposition, gas production, and waste stabilization. While there are significant economic advantages to the operation of landfills as bioreactors, our understanding of the design and operational practices of bioreactor technology and its impact on the environment is limited. When the bioreactor landfill reaches its permitted disposal capacity, the landfill must be closed and maintained under post-closure criteria, and there is still a need for a new landfill facility. That means we still have an open loop system of waste management. The limitations of conventional and bioreactor landfills can be summarized as

- Conventional landfilling practices require a long period of degradation.
- Bioreactor landfilling operations require less time than conventional landfilling practices.
- Both require a large space for operation.
- Both need to be closed after their operations have been completed.
- Once closed, there is need for space for new landfills.
- This is still an open loop waste management system.

For over-populated regions like Asia, Africa, and Latin America, and even for Europe, available space for new landfills is a major issue. In North America, regulations require post-closure activities and financial assurance for 30 years after landfill closure, and a state agency may require additional years of care, if needed (US EPA 2001). To address the problems related to space, a variation of the bioreactor, the "sustainable biocell" or "biocell" was first proposed by Hettiaratchi (2007). The biocell is frequently referred to as the third generation of landfills. It differs from traditional landfills by operating as a temporary storage facility, rather than a permanent one (Rahman 2018), and the space is reused (Hettiaratchi 2007), thus eliminating the need for post-closure monitoring. The biocell operation also accelerates waste decomposition and makes landfilling a treatment system for producing greater levels of renewable energy quickly, instead of serving as a perpetual storage facility. Hettiaratchi (2007) focused on accelerating the degradation of non-degradable or slow-degradable components and space recovery. The major differences between Hettiaratchi (2007) and the biocell operation research at UTA are:

- Hettiaratchi (2007) focused on accelerated decomposition of non-degradable components.
- UTA's objective was to operate the biocell as part of a solid waste management system and focus on the repeated reuse of space by accelerating degradation of most waste components for gas production and mining/digging out the leftover components for reuse or for composting or construction materials.

UTA's research focused on minimizing or eliminating plastics and other non-degradable/recyclable components from the waste stream from the beginning.

The major difference between the operation of bioreactor landfills and biocells:

- Bioreactor landfill – Only water or leachate is added and recirculated.
- Biocell – Both water/leachate and nutrients are added to accelerate decomposition.

Summary of the three landfill systems are shown below (Table 4.9).

The University of Texas at Arlington (UTA) research team demonstrated the feasibility and potential benefits of biocell operations both in the laboratory and the field. In both cases, the nutrients used were inexpensive and readily available microorganisms so they could be easily adapted by developing countries. A brief summary of the results of both research projects is presented below.

4.4.2.1 UTA Research on Gas Production: Laboratory-Scale Simulated Biocell Study

4.4.2.1.1 Background

The biocell operation was simulated using laboratory scale reactors to evaluate the effects of nutrients such as manure and sludge from the organic fraction of municipal solid waste (MSW) on methane production. MSW samples were collected from the working face of the City of Denton landfill in March (2016). Manure from cows, pigs,

Table 4.9 Comparison among three landfill systems.

	Traditional landfill	Bioreactor landfill	Biocell
Waste type	Mixed waste	Mixed waste	Mixed waste without Recyclables
Decomposition period	30–100 years	15–30 years	7–12 years
Gas Production rate	Very Slow	Moderate	Very Fast
Landfill-gas-to-energy	Not Viable due to very slow rate of gas production	Viable; however, the operation and moisture monitoring are very important	Viable
Addition or recirculation during operation	None – Target is to minimize moisture presence at landfill	Target is to increase moisture content with MSW waste – Moisture or leachate addition and recirculation	Both nutrients and moisture or leachate addition and recirculation
Space	Full	Full	Reuse space

and horses was used for this experiment because its high nitrogen ratio, low number of acidic bacteria, and increased hydraulic retention time of aged manure made it ideal for waste decomposition (Yazdani 2010). The physical composition of the MSW samples was determined by a wet weight basis. The MSW was sorted manually and categorized into paper, plastics, textiles, food waste, Styrofoam and sponge waste, metals, glass, yard and wood waste, construction debris, and others. The percentage of paper, food waste, textiles and leather, and yard and wood waste was fixed according to the physical composition of the degradable waste, and 50% paper, 20% food waste, 15% textiles, and 15% yard and wood waste were selected for the reactors.

4.4.2.1.2 Waste Characteristics

The MSW sample was composed of 34% paper, 19% plastic, 13% food waste, 8% textiles, 2% Styrofoam and sponge, 9% yard and wood waste, 3% metals, 2% glass, 4% construction debris, and 6% others (soils and fines). Results obtained from the physical composition test of MSW samples are presented in Figure 4.57.

Moisture content is the ratio of "pore" or "free" water to the dry or wet solid waste in a given mass of soil. It is expressed as a percentage. The moisture content is determined on both dry and wet weight bases; however, the moisture content for MSW is expressed as wet weight basis only. During the physical composition tests, the moisture content of the fresh waste samples was determined on both wet weight and dry weight bases. The initial moisture content (wet weight basis) for each reactor is listed in Table 4.10. The average moisture content of the fresh MSW in the MSW reactors was 26.7% on wet weight basis. The moisture content in the reactors was influenced by the moisture content of the waste feedstock, the moisture content of the inoculum, and the addition of water during the filling and compaction of the feedstock in the reactors. Though similar types of feedstocks were used in the

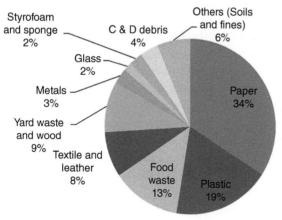

Figure 4.57 Average physical composition of MSW.

Table 4.10 Moisture content of MSW.

Reactors	Wet wt. (lb) in the reactor	Moisture content (%) (wet weight Basis)	Dry wt. (lb)	Total solid, TS (%)
M1, M2	2	26.61	1.47	73.39
M3, M4	2	26.58	1.47	73.42
M5, M6	2	25.70	1.49	74.3
M7, M8	2	27.32	1.45	72.68
M9, M10	2	27.43	1.45	72.57

Note: 1 lb = 0.45 kg.

reactors, due to the presence of different kinds of manure (cow, horse, and pig), the moisture content in the reactors varied to some extent. The additional water also ensured proper compaction of the waste and likely created an ideal ambient environment for proper microbial activity. The MSW reactors had an average of 73.27% total solids.

According to the Interstate Technology and Regulatory Council (2006), a volatile solid test is the most inexpensive measurement of the amount of biodegradable material that remains in a waste mass. The volatile organic content indicates the number of organic materials in the waste mass. The volatile organic content tests were conducted twice on the feedstock samples: before the reactors were sealed (initial volatile solid content) and at the end of the study. The initial volatile solids of the MSW reactors are listed in Table 4.11. Volatile solids accounted for about 85.9% of the MSW feedstock. According to Taufiq (2010), the volatile organic content of fresh MSW is usually about 76.96%. The higher percentage of volatile solids in this study was due to the use of organic MSW (paper, food waste, textiles, and yard waste) in the reactors.

Simulated biocell landfill reactors were built in the laboratory to analyze the gas generation and the effect of enzymes and manure on the degradation of the organic fraction of the MSW. Five pairs of MSW reactors were built as landfill biocell simulators (Table 4.12).

The experiment was conducted in 10 reactors incubated under laboratory conditions at a mesophilic temperature of 37 °C (98.6 °F) to simulate the actual landfill condition. The reactors were one-gallon HDPE wide-mouth plastic buckets (United States Plastic Corporation, OH) modified for gas and leachate collection, and liquid addition and recirculation. Paper, food waste, textiles and leather, and yard and wood waste were separated from the MSW. The MSW samples were shredded into squares that were 3.81 cm by 3.81 cm (1.5 in. by 1.5 in.) before adding them to the reactors since shredding improves waste decomposition. The recommended size for particles for maximum gas production is one-fourth to one-fifth the diameter of the bucket. Sufficient moisture for microbial activities was provided by spraying water on each layer of filling. Proper compaction was also maintained throughout the process. The reactors were filled 2.54–3.81 cm (1–1.5 in.)

Table 4.11 Volatile organic content of MSW.

Reactors	Inoculum	Volatile solid, VS (%)	VS (lb)
M1, M2	Control	86.71	1.35
M3, M4	Cow manure, 6%	84.41	1.29
M5, M6	Pig manure, 6%	87.04	1.35
M7, M8	Horse manure, 6%	86.40	1.34
M9, M10	MnP enzyme, 0.00000213%	84.96	1.31

Note: 1 lb = 0.45 kg.

Table 4.12 Combinations of feedstock and inoculum for laboratory experiment.

Reactors	Waste type	Sludge	Manure and enzyme
M1, M2	Organic MSW, 90%	10%	Control
M3, M4	Organic MSW, 84%	10%	Cow manure, 6%
M5, M6	Organic MSW, 84%	10%	Pig manure, 6%
M7, M8	Organic MSW, 84%	10%	Horse manure, 6%
M9, M10	Organic MSW, 84%	10%	MnP enzyme, 0.00000213%

below the top level of the bucket to provide a large enough area for gas to escape through the gas collection outlet. The lids of the reactors filled with waste were then sealed with a double layer of sealant. After sealing, the whole reactor setup was kept in the environmental growth chamber at a temperature of 37 °C (98.6 °F). Figure 4.58 shows the schematic diagram for the reactor setup and the reactors inside the environmental growth chamber. Different combinations of sludge and manure containing two pounds of organic MSW to provide microorganisms and sufficient nutrients to enhance the degradation process and neutralize the acidic environment by acting as a buffer were fed to the reactors. Of the five pairs of reactors built for this study, one pair had MSW with 10% sludge, which acted as a control reactor. This control reactor actually simulated the bioreactor landfill condition, as leachate was recirculated on a regular basis. In a conventional landfill, no leachate is recirculated. Feedstock of MSW with 10% sludge and 6% of three types of manure (cow manure, pig manure, and horse manure) was prepared for three pairs of reactors. The remaining reactor had feedstock of MSW with 10% sludge and 0.00000214% of MnP (manganese peroxidase) mixed with an organic fraction of municipal solid waste.

The leachate and gas generation of the simulated lab scale reactors were monitored on regular basis. The volume, pH, COD, and BOD of the generated leachate, along with the leachate recirculation, were measured during the entire monitoring period. The gas monitoring program involved measurements of its composition, rate, and volume. The differences in moisture content and volatile solids before and after the degradation were also determined.

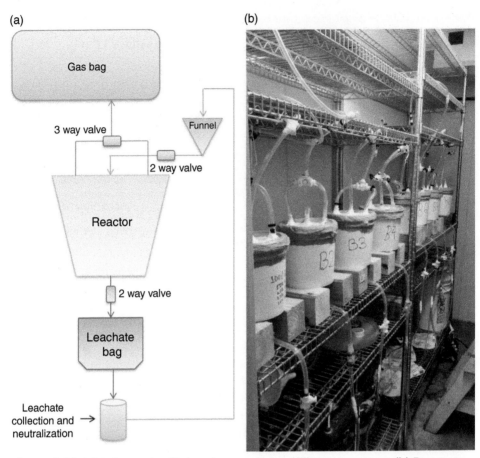

Figure 4.58 (a) Schematic of laboratory scale landfill reactor set up (b) Reactors set up in the laboratory.
Source: Courtesy of SWIS.

4.4.2.1.3 Gas Production Results

Landfill gas, or biogas generation, is the main indicator of waste degradation. In this study, the composition, volume, and rate of gas generated from the MSW reactors (M1–M10) were measured on a regular basis. Landfill gas mainly consists of methane, carbon dioxide, and oxygen, with traces of other gases such as nitrogen, hydrogen sulfide, ammonia, and non-methane organic compounds. The Landtec GEM 2000 was used in this study to measure the percentages of all of the different components of the gas generated in the reactors. After the reactors were installed, the first gas was measured on day 9 in MSW reactors M1–M10. At the beginning, the carbon dioxide content was high, and the methane content was low in all of the MSW reactors. The oxygen content was very low from the beginning due to the small size of the reactors; the amount of trapped oxygen during installation was also very low. As a result, the aerobic phase of the waste degradation process ended early, and the acidogenic phase commenced early. In the acidogenic phase, degradable organic compounds break into simpler compounds, such as carbon dioxide and water vapor. The carbon dioxide content peaked on day 9 and reduced with time, as the methane content rose. Other gases also decreased with time; however, the oxygen content remained steady at 2–3%. The methane content was below 2% for 20–25 days, and the pH of the leachate dropped below 5.5 in all of the MSW reactors due to the accumulation of volatile fatty acids (VFA). On the 25th day of operation, 10% of sludge was added to all of the MSW reactors during leachate recirculation, which helped to neutralize the acidic environment inside the reactors and start the methane production in reactors M5 and M6 with pig manure and reactors M9 and M10 with manganese peroxidase. Other MSW reactors started producing methane after 33–41 days. During the transition phase, between the acidogenic and methanogenic phases, the percentage of carbon dioxide reached 40% of the total gas composition. As soon as the methanogenic phase started, the methane-to-carbon-dioxide ratio ($CH_4:CO_2$) began expanding and the volume of other gases decreased. During the methanogenic phase, the methane content in all of the MSW reactors reached 60–65%, which was also seen in a study by Karanjekar (2013). The highest methane contents seen were 66.20 and 66.10% in the reactors with pig manure (M5 and M6, respectively), followed by the reactors with cow manure (M3 and M4). Reactors with MnP had methane contents of 64.1% in M9 and 62.6% in M10. The lowest methane content was seen in the control reactors, M1 and M2, which was 52.3 and 60.1%, respectively.

Figure 4.59 shows the variations of methane percentages with time for MSW reactors M1–M10. The highest methane contents seen were 66% in the reactors with pig manure (M5 and M6), followed by the reactors with cow manure (M3 and M4).

A cumulative gas generation graph is shown in Figure 4.60. Reactors with MnP (M9 and M10) produced the highest amount of gas in 233 days (100.6 l/lb and 105.1 l/lb, respectively). The reactors with pig manure (M5 and M6) produced about 85.4 l/lb and 83.3 l/lb, respectively, in 233 days, which was the second highest.

The cumulative methane-versus-time graph of MSW reactors (Figure 4.61) follows a trend similar to that of the gas-versus-time graphs. Reactors M9 and M10 generated about 54 l/lb (119.1 l/kg) and 52 l/lb (114.6 l/kg) of methane, respectively, in 233 days which was the highest among all of the MSW reactors and almost 22 times higher than the control reactor, M1. Reactors with pig manure (M5 and M6) also performed well in terms of methane production from the organic fraction of MSW, which was about 47 l/lb (103.6 l/kg) and 45.9 l/lb (101.2 l/kg), respectively, in 233 days. Although the reactors with MnP exhibited the maximum methane yield for the organic fraction of MSW, it is not recommended for field application, as it is expensive. The reactors with pig manure revealed that it can be as productive as the MnP, which is an advantage, as pig manure is readily obtainable at an affordable rate. Therefore, it is much more effective and economical to use pig manure as the additive for field scale biocell operations.

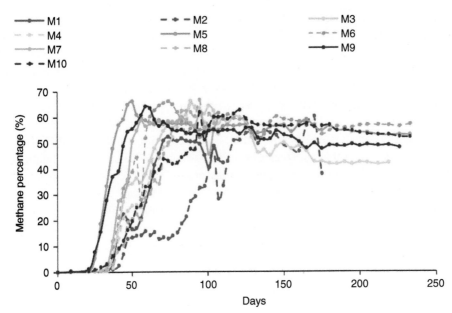

Figure 4.59 Methane content in MSW reactors

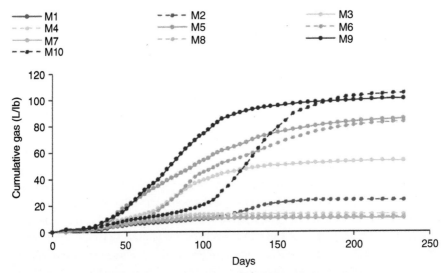

Figure 4.60 Cumulative gas generation (l/lb) in MSW reactors.
Note: 1 l/lb = 2.2 l/kg.

The rate of gas generation (gas yield) or methane generation (methane yield) is a cardinal indicator for landfill gas and follows a similar trend in all cases of MSW decomposition. The trend is that the gas or methane yield increases with time before it reaches a peak, then it decreases before ceasing. Figures 4.62 and 4.63 present the

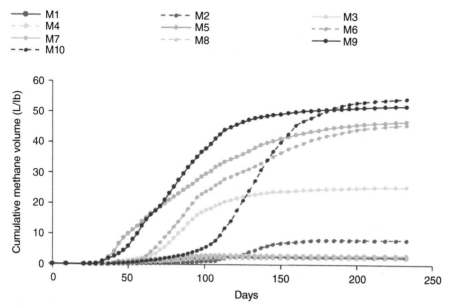

Figure 4.61 Cumulative methane generation (l/lb) in MSW reactors.

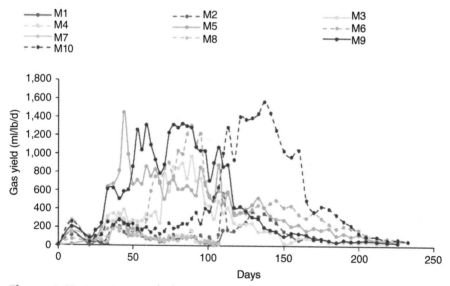

Figure 4.62 Gas yield (ml/lb/day) in MSW reactors.

rates of gas and methane generation of MSW reactors, respectively. One MSW reactor with pig manure (M5) achieved its peak on day 44, which was the earliest of all the MSW reactors and equates to about 1447 ml/lb/day (3190 ml/kg/day). Another MSW reactor with pig manure (M6) reached its peak on day 89, with 1304 ml/lb/day (2874 ml/kg/day).

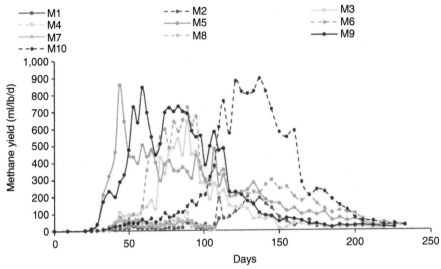

Figure 4.63 Methane yield (ml/lb/day) in MSW reactors.

Note: 1 ml/lb/day = 2.2 ml/kg/day.

Other experimental results can be found in the dissertation by Rahman (2018).

The rate of decomposition of organic material in the waste and the rate of landfill gas generation can be defined by the decay rate of waste. In the first order kinetic reaction, decay rate is defined as the biodegradation half-life of organic material (Half-life [$t_{1/2}$] is the time to break down 50% of the original amount of organic material in the waste). According to Barlaz et al. (1990), the rate of degradation of solid waste in landfills depends on the waste composition, waste particle size, moisture, ambient temperature, and pH. As the decay rate (k) value increases, the methane generation rate from landfills increases. In this study, the decay rate MnP and pig manure had the highest and second highest decay rates, respectively (Table 4.13).

Table 4.13 Decay rate of waste in MSW reactors.

MSW reactors	Decay rate, k (Yr⁻¹)
M1 (Control)	0.149
M2 (Control)	0.330
M3 (Cow manure)	1.015
M4 (Cow manure)	0.173
M5 (Pig manure)	**2.889**
M6 (Pig manure)	**2.707**
M7 (Horse manure)	0.188
M8 (Horse manure)	0.140
M9 (MnP)	**4.340**
M10 (MnP)	**5.859**

4.4.2.2 UTA Research on Gas Production: UTA Field-Scale Biocell Operation (Rahman 2018)

4.4.2.2.1 Background and Key Features

Based on the laboratory scale study, two test cells (control cell and biocell) were installed in the field. The cells were identical and contained the same feedstock, but nutrients were only applied in the biocell. The components of the cells included a leachate collection and removal system, a gas collection system, and a system to continually monitor the temperature and moisture of the MSW inside the cells. The key features of the field biocell are illustrated in the following section.

- The biocell was a custom-designed waste container made of steel, with dimensions of 6.4 m by 2.4 m by 2.4 m (21 ft. by 8 ft. by 8 ft.) It had a three-well head (Figure 4.64).

(a)

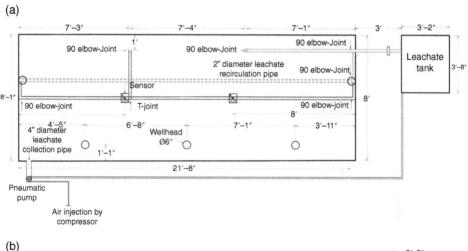

(b)

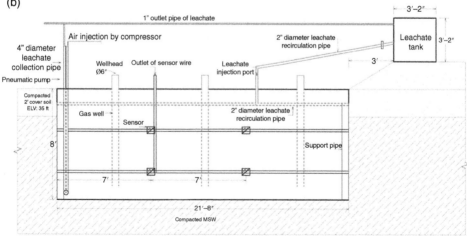

Figure 4.64 (a) Plan (b) Section of biocell.

- Each cell was equipped with three vertical perforated pipes for gas collection, a horizontal leachate recirculation pipe attached at two sides of the cell, and one leachate collection pipe.

- Three vertical gas wells were connected with a common 10.1 cm (4 in.) header pipe for gas collection. To measure the amount of methane production, a gas flow meter (ST100 Mass Flow Meter by FCI), was connected with the header pipe. The flow meter required a solar panel to supply continuous 24 Volt DC (direct currents).

- A layer of gravel covered the bottom of the cell, with a geotextile layer placed over it to provide for drainage of the leachate. The box was tilted on the slope to provide adequate gradient for the leachate to flow to the leachate collection sump pipe.

- Each cell was provided with a leachate reservoir tank. A pressure-activated pneumatic pump was used to extract leachate from the cell and send it to the reservoir. A VP4 bottom-loading pneumatic pump, manufactured by Viridian, was installed in the leachate collection sump. The pump required an air compressor to supply air to the pump. The leachate collection pipe was provided with a screener and a geotextile separation layer to avoid possible clogs in the pipe.

- Each cell had a lid with three 15.24 cm (6-in.) ports for gas collection, one 5.08 cm (2-in.) port for leachate recirculation, and another 5.08 cm (2-in.) port for removing the sensor cables from the cells.

- Four temperature sensors were installed in each cell and were supported by a two-layer PVC pipe frame that was attached at the side of the box. The sensor port was outside the PVC pipe so that it was in direct contact with the waste, while the sensor wire was inside the pipe and was connected with a data logger station.

The components of the cells included a leachate collection and removal system, a gas collection system, and a system to continually monitor the temperature and moisture of the MSW inside the cells. The biocell was a custom-designed waste container made of steel, with dimensions of 6.4 m by 2.4 m by 2.4 m (21 ft. by 8 ft. by 8 ft.). It had a three-well head (Figure 4.65). The container was designed with steel materials so that the temperature of the boundaries of the box was same as the surrounding ground. Each cell was equipped with three vertical perforated pipes for gas collection, a horizontal leachate recirculation pipe attached at two sides of the cell, and one leachate collection pipe.

Residential solid waste was chosen as the feedstock for the biocell, as it typically contains the high organic and moisture content that improves gas production. Based on the concept of biocells, only organic waste that had been sorted at the material recovery building at the City of Denton Landfill was deposited into the biocell. Approximately 5 tons of waste were deposited in the control cell, and 3.6 metric tons (4 US tons) of waste, 0.45 metric tons (0.5 US tons) of sludge (Class B type biosolids from the wastewater treatment plant), and 0.3 ton of pig manure were deposited in the biocell (Table 4.14). Waste was not compacted; rather was allowed to be compacted by its own weight.

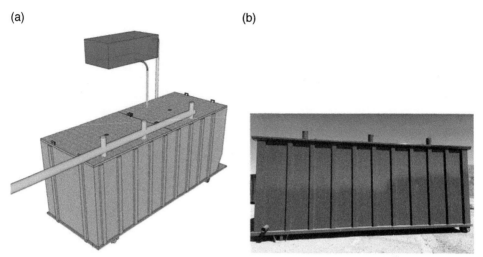

Figure 4.65 (a) Schematic of biocell (b) Image of biocell.
Source: Courtesy of SWIS.

Table 4.14 Combination of feedstock and inoculum for field scale.

	Feedstock (MSW)	Sludge	Pig manure
B1 (Control cell)	100%	–	–
B2 (Biocell)	84%	10%	6%

Similar to that which was used for the laboratory study, only the organic component of the municipal solid waste was utilized as the feedstock for the field-scale biocell. The recyclables and other materials had been removed during the processing, shown in Figure 4.67. The physical characteristics of the MSW utilized in the biocell construction and operation are presented in Figure 4.66.

The MSW was processed, and the organic components of the solid waste were separated at the Building Material Recovery (BMR) facility, located in the City of Denton landfill. Figure 4.67 shows the activities of waste sorting at the BMR facility at the City of Denton Landfill.

An extensive monitoring program was designed to collect data related to the leachate generation and recirculation, gas production, and waste degradation.

The construction and instrumentation activities began on 27 December 2016 and were completed on 23 January 2017. The field activities involved excavation; placing the steel box in the excavated ground; filling the cell with waste; installing leachate collection and recirculation systems, gas collection, and automated data collection systems; installing moisture and temperature sensors; and setting up a solar panel. Figures 4.68–4.72 present some of the activities of the field scale biocell installation.

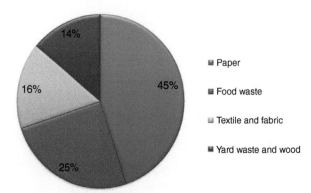

Figure 4.66 Physical composition of MSW.

Figure 4.67 (a) and (b) Waste sorting at Building Material Recovery (BMR) facility at City of Denton Landfill (c) and (d) Manual sorting of waste.
Source: Courtesy of SWIS.

(a)

(b)

Figure 4.68 Biocell installation (a) Excavation of trench. *Source*: Courtesy of SWIS. (b) Placement of steel box in excavated ground. *Source*: Courtesy of SWIS.

(a)   (b)

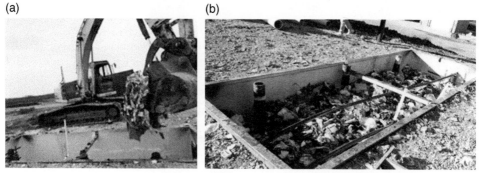

Figure 4.69 Waste placement in cells (a) waste being deposited into the steel box. *Source*: Courtesy of SWIS. (b) steel box filled up with waste. *Source*: Courtesy of SWIS.

Figure 4.70 Covered biocell connected with landfill gas-to-energy system and leachate recirculation system.
Source: Courtesy of SWIS.

(a)   (b)   (c)

Figure 4.71 Flowmeter installation.
Source: Courtesy of SWIS.

Figure 4.72 Solar panel to supply power to flowmeter.
Source: Courtesy of SWIS.

The leachate that was generated in both cells was collected and recirculated on a regular basis during the entire operation of the field study.

4.4.2.2.2 Analyses and Results of Field-Scale Study

The gas composition of the control cell and biocell are shown in Figures 4.73 and 4.74, respectively. In the middle of February 2017, the methane content in the control cell was only 8.47%. The biocell experienced more anaerobic activity than the control cell, based on the amount of carbon dioxide and methane. The biocell began producing methane earlier than the control cell, and the methane content reached almost 20% in the third month of monitoring (March 2017). The highest methane content in the control cell was about 45.4% in August 2017.

The methane content in the biocell peaked once during the last 14 months of monitoring (63.3%) and began to drop very slightly at the end of October 2017, showing that the methanogenic phase was longer in the biocell than in the control cell.

After 424 days of monitoring, about 30 336 standard cubic feet (SCF) of gas was produced from the biocell, with methane accounting for almost 12 437 SCF of it. After 425 days, about 15 553 SCF of gas was logged in the gas flow meter from the control cell, with methane accounting for only 4644 SCF of it. The cumulative volumes of gas and methane of the biocell and control cell are showed in Figures 4.75 and 4.76, respectively.

The highest gas flow, 22 standard cubic feet per minute (SCFM), was seen in the biocell; the average was about 16.7 SCFM. In the control cell, the highest gas flow rate was about 21 SCFM, and the average was about 8.9 SCFM, which was almost half of the average gas flow rate of the biocell. Figure 4.77 shows the gas flow rates of the biocell and control cell. Figure 4.78 shows the volume of methane generation throughout the monitoring period for the field-scale biocell and control cell.

Figure 4.79 shows the total methane yield of the field-scale biocell and control cell. The volume was converted to m^3/Wet Mg for comparison with previous studies. About 117 m^3 of methane was produced from 1 Mg of wet waste in 424 days in the biocell,

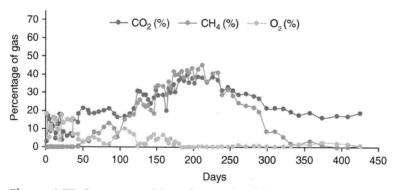

Figure 4.73 Gas composition of control cell (Day 1:5 January 2017).

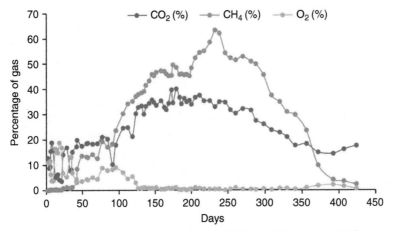

Figure 4.74 Gas composition of biocell (Day 1:7 January 2017).

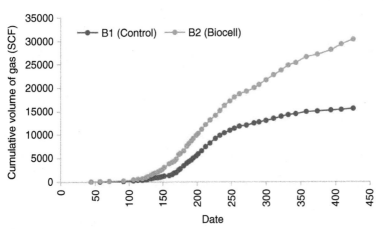

Figure 4.75 Cumulative volume of gas generated from control cell (Day 1:5 January 2017) and biocell (Day 1:7 January 2017).

whereas about $29\,m^3$/Wet Mg was produced from 1 Mg of wet waste in 425 days in the control cell. A study was conducted by Yazdani (2010) at the Yolo County Central Landfill in California, where a field-scale digester cell was operated in a two-stage batch system. The green waste was degraded under anaerobic conditions, followed by aerobic conditions, with aged horse manure serving as the inoculum. During the anaerobic phase of 451 days, about $60\,000\,m^3$ of methane was generated. The methane yield was $27.2\,m^3$ of methane per Mg of wet solid. From this comparison, it is comprehensible that the biocell produced almost four times more methane than the control cell.

The estimated decay rate of the biocell was considerably larger ($1.32\,year^{-1}$) than the decay rate of the control cell ($0.18\,year^{-1}$) and the values found in literature.

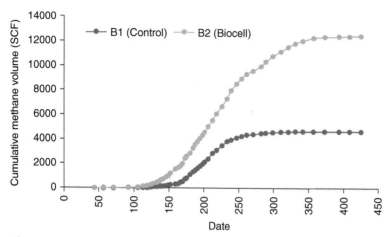

Figure 4.76 Cumulative volume of methane from control cell (Day 1:5 January 2017) and biocell (Day 1:7 January 2017).

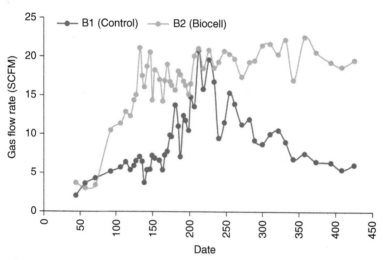

Figure 4.77 Flow rate (SCFM) of gas in controlled cell (Day 1:5 January 2017) and biocell (Day 1:7 January 2017).

Table 4.15 shows a comparison of the various decay rate constant values found in the literature with those from this study. The k values found in literature for bulk waste in U.S. landfills varied from 0.003 to 0.21 year⁻¹ (USEPA 2005). In a study by Hunte (2010), the kinetic rate constant for residential waste in Calgary was estimated to be 0.131 year⁻¹ and 0.146 year⁻¹ for the biocell he worked on. Yazdani (2010) conducted another study on green waste and horse manure in California and found the decay rate to be 0.82 year⁻¹.

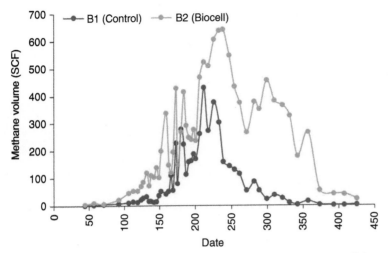

Figure 4.78 Volume of methane generation from control cell (Day 1:5 January 2017) and biocell (Day 1:7 January 2017).

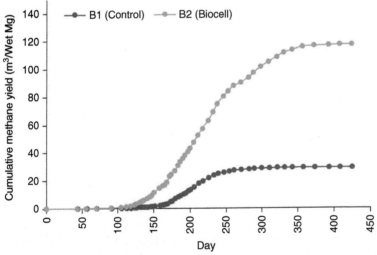

Figure 4.79 Cumulative methane yield in m³/Mg from control cell and biocell.

A higher value of decay rate from the field biocell in this study indicated accelerated waste decomposition. It was almost 11-fold and 33-fold the U.S. EPA bioreactor landfill default decay rate ($k = 0.12\,year^{-1}$) and conventional landfill default decay rate ($k = 0.04\,year^{-1}$), respectively.

The results from both the laboratory-scale and field-scale experiments were excellent. Therefore, it can be concluded that operating a landfill as a biocell will both accelerate and increase the gas production and finally result in higher

Table 4.15 Comparison of decay rate constants of current study with values reported in literature.

Decay rate constant	Type of landfill	References
$k = 0.02\,\text{yr}^{-1}$	Dry landfill receiving less than 20 in. of annual precipitation	EPA (2010)
$k = 0.04\,\text{yr}^{-1}$	Moderate landfill receiving between 20 and 40 in. of annual precipitation	EPA (2010)
$k = 0.06\,\text{yr}^{-1}$	Wet landfill receiving more than 40 in. of annual precipitation	EPA (2010)
$k = 0.12\,\text{yr}^{-1}$	Bioreactor landfill operating as bioreactor, where water is added until the moisture content reaches 40% moisture on a wet-weight basis	EPA (2010), Barlaz et al. (2010) and Tolaymat et al. (2010)
$k = 0.052\,\text{yr}^{-1}$	National average, corresponding to a weighted average based on the share of waste received at each landfill type	EPA (2010)
$k = 0.3\,\text{yr}^{-1}$	Proposed for well-designed wet landfills in the U.S.	Faour et al. (2007)
$k = 0.146\,\text{yr}^{-1}$	Calgary biocell	Hunte (2010)
$k = 0.82\,\text{yr}^{-1}$	Estimated for digester cell fed with green waste and horse manure	Yazdani (2010)
$k = 1.32\,\text{yr}^{-1}$	Biocell	Rahman (2018)
$k = 0.18\,\text{yr}^{-1}$	Control Cell	

energy generation. This is one of the major components of a sustainable waste management system.

4.4.2.2.3 Landfill Gas-to-Energy

Landfill gas is a prodigious alternative for fossil fuel for every generation. It can be used to produce a significant amount of electricity as it contains approximately 40–60% methane (CH_4), and carbon dioxide (CO_2).

The methane generation rate per year from the waste mass disposed in a landfill, can be calculated using the following formula developed by USEPA (1998):

$$Q_M = \sum 2.k.L_o.M_i\left(e^{-k.t_i}\right)$$

Where,

L_o = Methane generation potential in ft³/ton
M_i = Mass of solid waste place in landfill in the ith year in ton
k = Decay rate constant in year⁻¹
t_i = Age of the waste mass in the tth year in year

Assuming 50% collection efficiency of total produced methane, electricity generation can be estimated using the following formula:

$$Q_{electricity} = Q_{methane} \frac{ft^3}{year} \times 600 \frac{BTU}{ft^3} \times \frac{2.93 \times 10^{-4} kwh}{BTU} \times 0.5$$

1 kWh/year = 1.14079553 × 10^{-7} megawatts (MW)
Calorific value of 1 ft^3 biogas = 600 BTU
1 BTU = 2.93 × 10^{-4} kwh

4.4.3 Landfill Mining of Biocell Operation

Landfill Mining as a Part of Sustainable Resource Management: Conventional waste management is an open-loop cycle where generated products end up in the landfill after being utilized by consumers (Braungart and McDonough 2009). Problems associated with traditional landfills include loss of materials that still have value, air and water quality impacts, climate change impacts, and post-closure monitoring costs. These problems can be minimized via a sustainable resource management system (Figure 4.1), which consists of three main components: (i) material recovery, (ii) landfill operation as a biocell, and 3) landfill mining. These components increase material recovery and reuse of materials like paper, plastic, and metals, and provide quicker degradation and renewable energy from non-recoverable wastes like food and yard waste (via biocell operation, discussed below). Landfills will no longer be long-term storage facilities, but instead be treatment systems that can be operated perpetually in the same location. This should lead to improved public perception and acceptance by the greater urban community.

Landfill mining refers to the process of excavating previously disposed of MSW from a landfill for the purpose of recovering materials of value for re-use and reuse of the excavated land for future development. Landfill mining was first initiated in Tel Aviv, Israel in 1953 with the primary objective to recover the soil amendments from the excavation (Savage 1993). This was the only initiative reported for several decades, until the early 1990's. From 1995 to 2000, interest in landfill mining strategies grew, and several landfill mining projects were reported around the world (Cossu et al. 1996).

As part of the Sustainable Resource Management System, landfill space (volume) can be reutilized, and waste management operations can be converted to a "closed loop" system through landfill mining (as presented in Figure 4.80). In the generic closed-loop system depicted in Figure 4.80, alternative waste technology refers to a substitute for traditional landfilling. Such alternative technologies recover valuable materials or energy from waste and include materials recovery/recycling facilities, landfill mining to recover recyclable materials, composting of food and yard waste, and waste-to-energy plants. The major benefits of landfill mining in a closed loop system are conservation

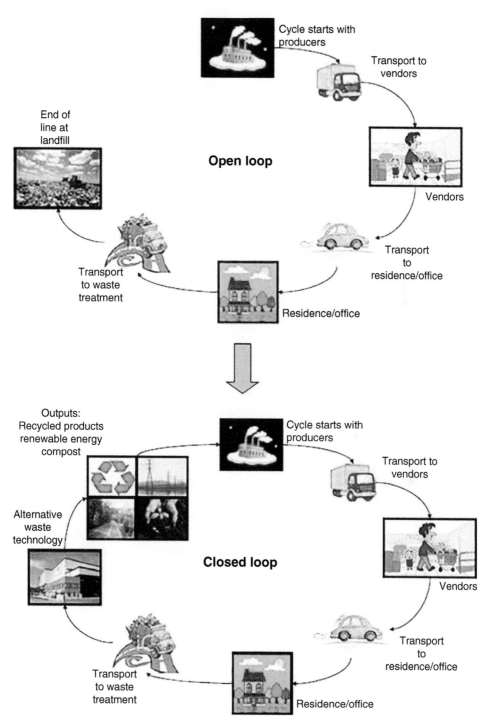

Figure 4.80 Open loop to closed loop operation of waste management through landfill mining.

Source: Based on Hossain et al. (2014)/with permission of American society of civil engineers.

of landfill space and recovery of valuable recyclable materials (Hossain et al. 2014). The utilization of SMART Facility can help all these operations in one place and accelerate sustainable waste management in developing countries.

4.4.3.1 Feasibility Study of Landfill Mining in Texas

Landfill mining for material recovery and as a source of energy provides an excellent solution for numerous problems related to waste deposits. As waste materials in landfills are very heterogeneous, mining can be associated with high capital investments. Therefore, for the following reasons, it is very important to evaluate the feasibility of landfill mining for an existing landfill and to determine the composition and characteristics of the excavated material.

1. It helps determine the technical and economic feasibility of the mining operation and space recovery.

2. Mined waste can be recycled or reused based on the level of degradation and materials recovered during the mining operation.

3. Mined waste can be used as a source of energy by burning it in waste-to-energy plants. The waste-to-energy facilities need the energy potential of mined municipal solid waste as it appears to be one of the important parameters in the design of incinerators (Xu et al. 2015).

4. According to Lee and Jones-Lee (1999), another issue that is not incorporated into evaluation of landfill gas production is that much of the municipal solid waste placed in landfills is deposited inside polyethylene bags. These bags, while crushed, are not shredded, and may act as barriers to moisture interacting with the components within the bags. This may inhibit the fermentation of the organics in bagged wastes due to low moisture content. The plastic bags decompose slowly and, even though the duration of the integrity of the polyethylene plastic bags is unknown, it is likely to be on the order of at least decades. The net result is that the production of landfill gas can potentially take place over many decades and may extend to hundreds of years. If the solid wastes are not degraded, there is tremendous potential for landfill mining that may offer economic benefits for landfill owners. Therefore, it is important to determine the physical and engineering characteristics, as well as the state of decomposition, of landfilled municipal solid waste.

5. Many MSW landfills in the United States were constructed prior to the current federal regulations and were typically unlined, with minimal or no leachate collection systems. These landfills, many of which are closed now, might be a significant source of leachate and landfill gas (LFG). During the lifetime of the final cover, there is a possibility of deterioration of the final cover. This may allow water to get into the landfill, and with the presence of moisture, the landfill may start producing gas again. Lee and Jones-Lee (1999) discussed that once a

landfill is closed and the low-permeability landfill cover is installed, the rate of moisture entering the landfill is very low and gas production may decrease due to absence of water or moisture. However, even decades after a landfill has been closed, the unfermented organic components of the waste can again initiate gas production proportional to the moisture content of the waste.

The authors conducted a feasibility study of the landfill mining operations of the closed sections of two landfills in North Texas, USA. Both of these feasibility studies were conducted: (i) to implement biocell operation as sustainable waste management operation and to reuse the mined cell for a future biocell operation (as discussed previously), (ii) to study the characteristics and potential use of mined waste materials (including energy potential for using them in WTE facility or other facilities as needed), and (iii) to study feasibility of landfill mining as part of circular economy. The following sections describe the methodology and present the results from the studies.

4.4.3.2 Case 1 - City of Denton Landfill in Texas, USA

A study was undertaken to investigate the state of decomposition of the landfilled waste and to evaluate the feasibility of landfill mining for a closed section of the City of Denton Landfill in Texas, USA. The solid waste was collected, the physical composition of the waste samples was determined, and the moisture content, unit weight, hydraulic conductivity, particle size distribution, and volatile solids' fraction of the collected samples were defined.

4.4.3.2.1 Background

The study area was located in a City of Denton Landfill cell that is located on the southeast side of Denton, Texas and began operating in 1983. The Denton landfill received its permit to start accepting waste on 7 March 1983 (permit number of 1590) and Cell 1590, later modified as Cell 1590 A, was pre-subtitle-D. Initially the landfill area was 32 acres, but in 1998, it was expanded to cover 252 acres (0.39 sq-mile), with 152 acres (0.24 sq-mile) for solid waste and 100 acres (0.17 sq-acres) for offices, a buffer zone, compost, and extra rented land. At the time of the study there were six cells in the landfill and the former cell was considered as cell zero or cell 1590 A. The landfill then received approximately 550 tons (1.1 million lb) of MSW a day. It is classified as a Type I landfill, which means that it is a standard landfill for the disposal of municipal solid waste (MSW). The landfill follows operational rules cited in the 30 TAC 330 subchapters D, provided by the Texas Administration Code. In 2009, the landfill transitioned to an enhanced leachate recirculation landfill to increase the gas production and to increase capacity of the landfill space by improved waste degradation. For the presented study, solid waste samples were collected from two boreholes B-70 and B-72 of cell 1590 A, as presented in Figure 4.81.

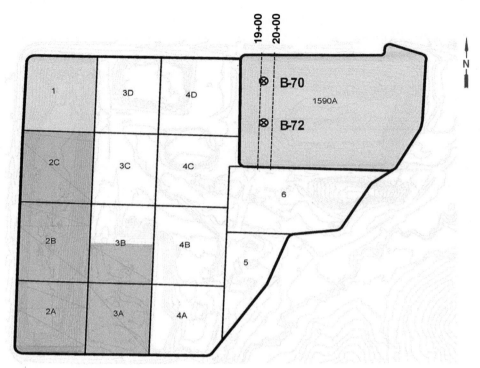

Figure 4.81 Location of boreholes for the collection of waste samples.

4.4.3.2.2 Mined MSW Sample Collection

MSW samples were collected from the Denton landfill in November 2010. A 0.91 m (3 ft.) diameter bucket augur attached to an AF130 hydraulic drill rig was used for drilling, as shown in Figure 4.82. Solid waste samples, some of which were as old as 25 years, were collected from two boreholes (B-70 and B-72). The age of the samples was estimated from the dated newspapers and magazines that were in the collected samples, and from the history of the landfill operation. Six samples were collected from each borehole starting at 3.05 m (10 ft.) of depth and then at 3.05 m (10 ft.) intervals, up to 18.3 m (60 ft.). Based on the findings from a previous study on the adequate weight of samples required for characterization (Taufiq 2010), 11.3–13.6 kg (25–30 lb) of MSW were collected at each depth.

The collected samples were taken to the laboratory in closed plastic buckets and were stored and preserved at approximately 38 °F (below 4 °C) in an environmental growth chamber.

4.4.3.2.3 Physical Composition

The physical composition of MSW varies within a landfill. The degradable constituents in the MSW decrease with time and decomposition, thereby resulting in variations in

(a)

(b)

(c)

Figure 4.82 Sample collection at the City of Denton Landfill (a) Hydraulic drill rig (b) 3-ft. diameter Bucket Auger (c) Sample collection.
Source: Courtesy of SWIS.

the overall composition. The composition of the MSW from boring B-70 is presented in Figure 4.83 as the percentage of weight of individual waste components to the weight of the total waste.

From a visual inspection of B-70, it was determined that the top 2.4 m (8 ft.) of the boring was cover soil. MSW samples collected from a 3.05 m (10 ft.) depth were relatively fresh, samples collected from 6.10 m (20 ft.) and 9.14 m (30-ft.) depths were partially degraded and the rest of the samples (deeper than 9.14 m) were mostly non-degraded. The closed landfill section was operated as a conventional landfill; therefore, no water was added to the landfill. No permanent cover was provided, except for cover soil on top of the landfill. The higher degradation at 6.10 m (20 ft.) and 9.14 m (30 ft.) might be due to water intrusion from the top through the cover soil. Figures 4.84 and 4.85 show the samples collected from B-70 at 6.10–9.14 m (20–30 ft.) and 12.2–18.3 m (40–60 ft.) depths, respectively.

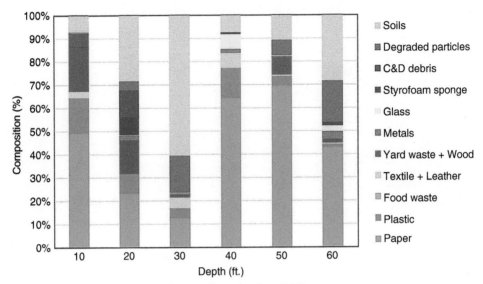

Figure 4.83 Composition of MSW from boring B-70.

Figure 4.84 Degraded samples from Boring B-70 at (a) 6.10 m (20 ft.) (b) 9.14 m (30 ft.) depths.
Source: Courtesy of SWIS.

The degradation of landfilled samples is conventionally anticipated to occur with age and depth; however, the collected samples did not follow this trend, as illustrated in Figure 4.83. Since moisture is not added to a traditional landfill, the moisture content of the waste varies with its composition. However, the absence of a final cover might lead to unanticipated water intrusion in the waste mass. There was presence of less-degraded samples taken from B-70 after 9.14 m (30 ft.) depth. In summary, the unavailability of moisture to the landfilled waste may result in less degradation or no degradation of the waste. No food waste was obtained from any of the samples

(a)

(b)

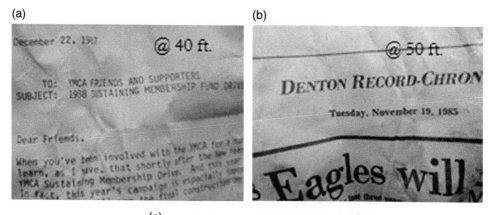

(c)

Figure 4.85 Samples from Boring B-70 at (a) 12.2 m (40 ft.) (b) 15.24 m (50 ft.) (c) 18.3 m (60 ft.) depths.
Source: Courtesy of SWIS.

from boring B-70. The paper content increased after 9.14 m (30 ft.) depth, where the collected samples were relatively fresh and, as presented in Figure 4.85, the dates of the papers were still legible and dated back to 1985. The presence of higher paper content indicates that the landfilled waste had not degraded. Therefore, it can be stated that a large portion of degradable waste was available during the time of the sample collection.

The landfill cover was approximately 4.57 m (15 ft.) for B-72; therefore, the sample collected from 3.05 m (10 ft.) depth was cover soil and was discarded. The composition of the MSW from boring B-72 is presented in Figure 4.86. No food waste was present in any of the samples. At 15.24 m (50-ft.) depth, the percentage of construction and demolition debris (C & D) was approximately 35%, which is very high. The samples from B-72 were more degraded than those from B-70, which might be attributed to an uneven distribution of moisture in the landfill and the extensively heterogeneous nature of waste. As degradation is dependent on the availability of moisture, it can be predicted that B-72 was in a more saturated zone than B-70.

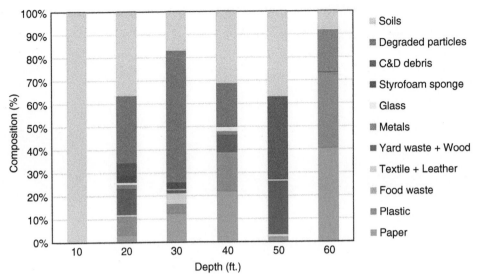

Figure 4.86 Composition of MSW from Boring B-72.

(a)

(b)

Figure 4.87 Degraded samples from Boring B-72 at (a) 6.10 m (20 ft.), (b) 15.24 m (50 ft.) depths.
Source: Courtesy of SWIS.

From visual inspection, it was determined that the samples collected from 6.10 m (20 ft.) and 15.24 m (50-ft.) depths were mostly degraded and black in color, as presented in Figure 4.87. The MSW samples from these two particular depths contained mostly fine components. The samples collected from 9.14 m (30 ft.) and 12.19 m (40 ft.) depths were partially degraded as presented in Figure 4.88, while the sample from 18.3 m (60 ft.) depth was less degraded, as shown in Figure 4.89.

The average composition of the landfilled samples, including both borings, is presented in Figure 4.90. The major component of the landfilled waste was determined to be paper

(a)   (b)

Figure 4.88 Partially degraded samples from Boring B-72 at (a) 9.14 m (30 ft.), (b) 12.19 m (40 ft.) depths.
Source: Courtesy of SWIS.

Figure 4.89 Samples with almost no degradation from 18.3 m (60 ft.) depth of Boring B-72.
Source: Courtesy of SWIS.

which was relatively high for landfilled waste (31%). Soils and degraded fines were determined to be 25%. Food waste was 0% and plastic was 10% in the total waste mass.

4.4.3.2.4 Comparison of Composition of Fresh and Landfilled Waste

MSW is highly heterogeneous and consists of many degradable components that are expected to degrade with time; albeit with varying degradation rates that are dependent upon the type of waste. Food waste is readily degradable, while paper, textiles, wood, and yard waste degrade at a slightly lower rate. Food waste may not be present in landfilled waste due to its fast degradation rate, while other degradable

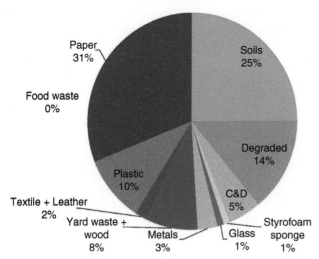

Figure 4.90 Average compositions of waste sample from the City of Denton Landfill.

waste will have a reduced mass. In contrast, the mass of non-degradable components remains the same in the MSW. Therefore, the percentage of degradable components may be less in degraded waste than in fresh waste. The percentage of non-degradable components may even increase. The average composition of the landfilled waste collected from the closed section of the landfill was compared to the annual average of fresh MSW from the working phase of City of Denton Landfill (Taufiq 2010), and the results are illustrated in Figure 4.91.

As expected, the percentage of paper in the landfilled waste was less than that of fresh waste, due to the degradation of the waste. The percentage of plastic waste was also lower than that of fresh. The landfilled waste was approximately 25 years old and plastic packaging was not widely used until 1990's, which might explain its lower percentage. The soils and fines were higher in the landfilled waste than in the fresh MSW, which might be attributed to the presence of degraded fines and lot of cover soil that was absent in fresh waste samples.

4.4.3.2.5 Moisture Content

The moisture content of MSW is extremely important, as it influences the decomposition behavior and all of the other engineering properties. The moisture contents of B-70 and B-72 on dry weight and wet weight basis are presented in Figure 4.92. The moisture contents (wet weight basis) of the samples from borings B-70 and B-72 averaged 28.82% and 20.27%, respectively. The average moisture content of the B-72 samples was lower than that of B-70 samples. According to Landva and Clark (1990), high organic content in MSW increases the moisture content, and the

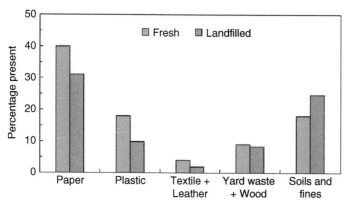

Figure 4.91 Comparison of fresh and landfilled waste.

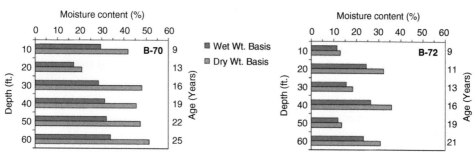

Figure 4.92 Moisture contents of MSW samples from Borings B-70 and B-72.

organic components decrease with degradation. Therefore, it might be concluded that the moisture content decreases with degradation. The samples from B-72 were more degraded than those from B-70, which might be the main reason for the low moisture content of the samples from B-72. Furthermore, the higher percentage of paper in the B-70 samples might be responsible for the higher moisture content.

The moisture contents of the fresh and landfilled waste were also compared and showed decreased moisture content in the landfilled waste due to decomposition. (See Figure 4.93.)

4.4.3.2.6 Volatile Solids

According to the Interstate Technology and Regulatory Council (ITRC) in 2006, measurement of volatile solids is the most inexpensive method of determining the amount of biodegradable material that remains in the waste mass. The volatile solids of landfilled MSW are linearly correlated with cellulose and cellulose/lignin data (Ham and Barlaz 1987), and a correlation between methane potential (BMP test results) and volatile solids was developed by Kelly et al. (2006). An increasing trend of volatile solids with

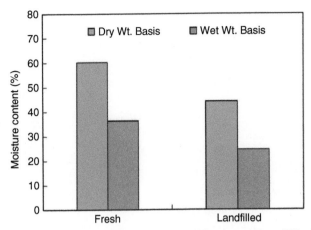

Figure 4.93 Moisture contents of fresh and landfilled waste.

depth was observed from the test results. The waste samples collected from B-70 and B-72 were mostly degraded on the top and less degraded on the bottom of the landfill. The average volatile solids in the samples from B-70 and B-72 was 72.29 and 61.44%, respectively. The sample collected from a 3.05 m (10 ft.) depth of B-72 was discarded, as it was only cover soil. The average for B-72 was less than the average for B-70. The samples collected from B-72 were mostly degraded, which is reflected in the lower percentage of volatile solids. The volatile solids for B-70 and B-72 are presented in Figure 4.94.

4.4.3.2.7 Comparison of Volatile Solids of Fresh and Landfilled Waste

The annual average percentage of the volatile solids in the fresh solid waste collected from the working face of the City of Denton landfill (2009–2010) was determined as 76.96%.

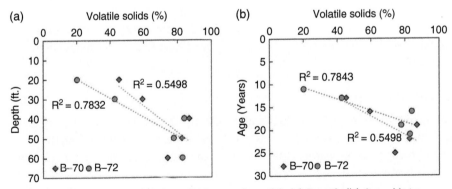

Figure 4.94 Volatile solids in MSW samples with (a) Depth (b) Age. *Note:* 1 ft. = 0.3048 m.

The average percentage of volatile solids of the landfilled waste after discarding the cover soil sample was 66.87%. The percentage of volatile solids in the landfilled waste was only slightly less than that in the fresh waste, which indicates that the samples were only partially degraded and there is a probability of future gas generation from these samples.

In general, gas production is a function of the waste composition. Cellulose and hemicelluloses comprise 45–60% of the dry weight of MSW and are its major biodegradable constituents (Barlaz et al. 1989). The decomposition of these compounds produces methane (CH_4) and carbon dioxide (CO_2) in landfills (Barlaz et al. 1990).

Kelly et al. (2006) developed correlations between cellulose and VS, lignin and VS, BMP and VS, and cellulose + lignin and VS. Based on the correlations and test results, BMP, cellulose, lignin and C/L were estimated, as illustrated in Tables 4.16 and 4.17, and

Table 4.16 Predicted BMP, cellulose, and lignin percentage from volatile solid results for B-70.

Depth	Age	VS	BMP	Cellulose (C)	Lignin (L)		Expected phase of
(Ft.)	(Years)	(%)	(ml/g)	(%)	(%)	C/L	degradation
10	9	85.79	134.67	41.36	17.50	2.36	Fresh sample
20	13	45.30	70.69	20.31	20.86	0.97	Between phase I and II
30	16	59.28	92.78	27.58	27.47	1.00	Between phase I and II
40	19	86.84	136.33	41.91	26.45	1.58	Phase I
50	22	82.57	129.58	39.69	24.38	1.63	Phase I
60	25	73.95	115.96	35.20	23.98	1.47	Phase I
	Average	72.29	113.34	34.34	23.44	1.47	

Table 4.17 Predicted percentage of BMP, cellulose, and lignin from volatile solids in B-72.

Depth	Age	VS	BMP	Cellulose (C)	Lignin (L)		Expected phase of
(Ft.)	(Years)	(%)	(ml/g)	(%)	(%)	C/L	degradation
10	9	5.00					Cover soil
20	11	20.20	31.04	7.25	16.86	0.43	Between phase II and III
30	13	42.61	66.44	18.91	26.77	0.71	Between phase II and III
40	16	83.90	131.68	40.38	25.28	1.60	Phase I
50	19	77.69	121.87	37.15	26.50	1.40	Phase I
60	21	82.80	129.94	39.81	21.38	1.86	Phase I
	Average	61.44	96.20	28.70	23.36	1.23	

Figure 4.95. The C/L ratio was less than or equal to 1 for samples from 6.096 m (20 ft.) and 9.144 m (30 ft.) of B-70 and B-72. The C/L ratio for the sample from 3.048 m (10 ft.) depth of B-70 was 2.36. The C/L ratio or (C+H)/L ratio is directly related to degradation of waste. Hossain et al. (2003) presented a correlation between the (C+H)/L ratio and the compression index that is directly proportional to the settlement of waste. Therefore, the C/L ratio provides a good indication of the decomposition level of waste.

The C/L ratio increased with depth for borings B-70 and B-72. Therefore, it can be summarized that the landfilled samples were more degraded on the top and less degraded on the bottom. According to Hossain et al. (2003), the (C+H)/L ratio decreases with the degradation of landfilled wastes. The (C+H)/L ratios for different phases of the samples are illustrated in Figure 4.96. The (C+H)/L ratio for the bioreactor samples

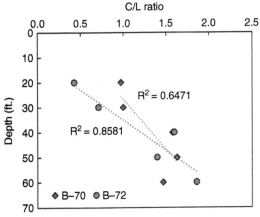

Figure 4.95 C/L ratio with depth for Borings B-70 and B-72.

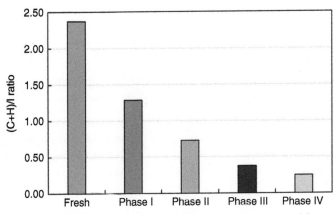

Figure 4.96 (C+H)/L ratio with phases of decomposition of MSW (Hossain et al. 2003).

decreased concurrently with the increasing volume of methane produced; the decrease in this ratio in the samples from the traditional landfill was slower. The (C+H)/L ratios of the traditional samples were higher than those of the bioreactor samples at the same time point. It was documented that when the bioreactor samples had completed all of the degradation phases, the traditional samples were still in early Phase III.

The closed section of the landfill was operated as a traditional landfill and the C/L ratio was more than 1 for most of the samples, indicating that the degradation of the landfilled waste was not complete at the time of the sample collection. Based on a previous study (Hossain et al. 2002), the degradation phases of the samples were determined from the estimated C/L ratio. The C/L ratio indicated that the MSW samples collected from the Denton Landfill were between Phase I and IV.

4.4.3.3 Case 2 – City of Irving Landfill in Texas, USA

A study was conducted to characterize the landfilled solid waste and determine its state of decomposition to predict the energy potential of the mined waste, and potential reuse as a mined section of the landfill for future biocell operation. The waste samples were collected from a closed section of the City of Irving Landfill in Texas, USA. The authors collected samples of the landfilled solid waste and determined its physical composition, moisture content, volatile solids' fraction, and calorific value.

4.4.3.3.1 Background

The City of Irving Landfill, located on the south side of Irving, Texas, received permission to start operation in December of 1980 and actually began operating in April of 1981. It serves more than 250 000 residents under the direction of the city's Solid Waste Service Department (SWS), who is the exclusive service provider for residential waste collection. The SWS owns and operates the Type 1 Hunter Ferrell Landfill, accepting residential waste from the city and commercial waste from a designated waste hauler. Approximately 800 tons of waste are sent to the landfill every day. The waste footprint for the landfill is 139.5 acres (0.22 sq-mile), and the disposal capacity of the landfill is 16 277 722 m³ (21 290 457 cubic yards). The landfill is divided into tracts identified by the City as east, west, and middle tracts. After nearly 40 years of operation, the west tract was nearing maximum capacity. Currently, the middle tract is being used to expand the waste disposal capabilities for continuous operation. Waste samples were collected from three boreholes (X, Y, and Z), in the west tract that are shown in Figure 4.97. Borehole X was at the initial location of the landfill that accepted waste from 1982 to 1992.

The landfill waste samples were collected from the closed cells of the landfill in May 2016. An AF130 hydraulic drill rig with a 0.91 m (3 ft.) diameter bucket augur were used for drilling, as shown in Figure 4.98. Samples were retrieved at every

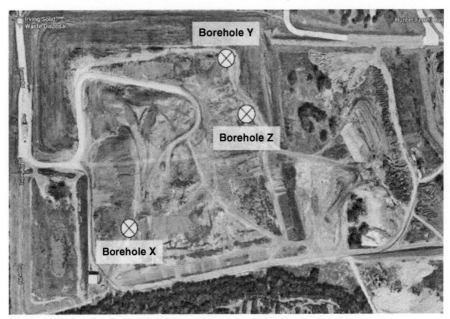

Figure 4.97 Location of boreholes for the collection of waste samples.

3.05 m (10 ft.) depth in each borehole. For the X, Y, and Z boreholes, 15.24 m (50 ft.), 21.3 m (70 ft.), and 30.48 m (100 ft.) boreholes, respectively, were dug. Based on the dated newspapers and magazines retrieved, as well as from the landfill operational history, the collected waste samples were estimated to be as old as 35 years. The collected samples were brought to the laboratory in sealed plastic bags, and all of the bagged samples were stored and preserved at approximately 38 °F (below 4 °C) in an environmental growth chamber.

A total of 22 mixed waste samples were obtained for the analysis. Four components of the mixed waste (plastic, paper, textile, and wood) were considered for the energy calculation. A total of 88 samples were tested for their calorific value.

4.4.3.3.2 Physical Composition

The composition of the MSW from Borehole X is presented in Figure 4.99 as a percentage of weight of the individual waste components to the total weight of the waste. The degradable fraction was higher in the first 9.144 m (30 ft.) From a visual inspection, the excavated waste appeared to be relatively fresh, as it was disposed of one or two years ago in a cell that was 25–35 years old. The paper fraction gradually decreased from 25.34 to 5.20% up to 12.19 m (40 ft.). The plastic fraction was 22.63% at 6.096 m (20 ft.) depth, which is higher than average; the fine fraction varied between 37 and 54%. The average composition of Borehole X was 14% paper, 11% plastic, 12% textile + leather, 11% yard and wood waste, 3% metals, 1% glass, 2% C&D debris, and 46% others (mixed other objects and fines).

(a)

(b)

(c)

Figure 4.98 Sample collection at the City of Irving Landfill (a) Hydraulic Drill Rig (b) 3 ft. diameter Bucket Augur (c) Sample collection.
Source: Courtesy of SWIS.

The top 6.096 m (20 ft.) of the landfill had relatively fresh waste, but below this depth, the waste was around 35 years old. The average fine content across the depth of the borehole was 46%. No food waste was observed. The non-combustible portion of waste was around 54% while the combustible portion was around 46%. As the depth and age of waste increases, the waste is expected to be more degraded; however, as this is a conventional landfill, water was not added to the fill, so, there was no uniform degradation. Figure 4.100 shows a paper from 1985 that was found at 9.144 m (30 ft.) and had very little degradation.

The composition of the MSW from Borehole Y is presented Figure 4.101. It had about 66% fine content which indicated the highest amount of degradation among all boreholes. There was much less moisture, as was observed during the drilling, and the remaining environment was very dry and unsuitable for further degradation. No food

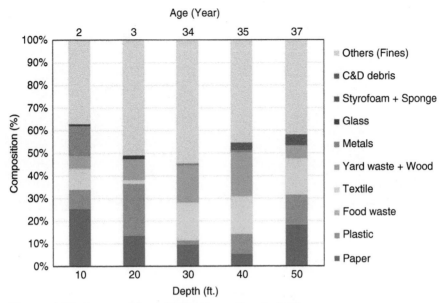

Figure 4.99 Composition of MSW from Borehole X.

Figure 4.100 (a) Waste sample from Borehole X (b) Paper found at 30 ft. depth.
Source: Courtesy of SWIS.

waste was found. The non-combustible portion of the waste collected was found to be 68%, and the combustible portion was about 32%. Similar to the other boreholes, no trend was found in the composition of the waste with age or depth of the fill. The paper waste samples from 1985 to 1997 were fresh and readable, as shown in Figure 4.102. The average composition of Borehole Y was 14% paper, 10% plastic, 2% textile + leather, 6% yard and wood waste, 1% metals, 1% C&D debris, and 66% others (mixed other

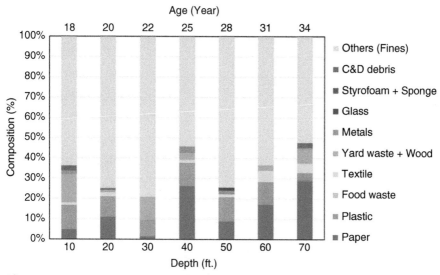

Figure 4.101 Composition of MSW from Borehole Y.

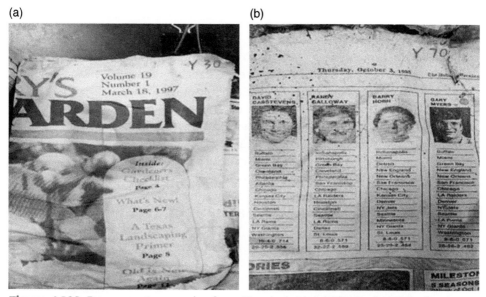

Figure 4.102 Paper waste samples from Borehole Y at (a) 30 ft., (b) 70 ft. depths. *Source*: Courtesy of SWIS.

objects and fines).

The composition of the MSW from Borehole Z is presented Figure 4.103. It had the highest moisture content of the boreholes, and no food waste was found. The non-combustible portion of the waste collected was found to be 65%, and the combustible portion was about 35%. Similar to the other boreholes, no trend was found in the

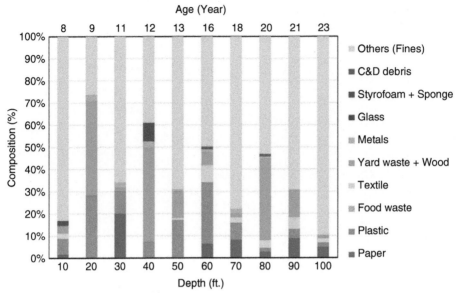

Figure 4.103 Composition of MSW from Borehole Z.

composition of the waste with age or depth of the fill. Degradation is a function of moisture availability; therefore, it can be predicted that the waste present in the vicinity of Borehole Z was in a more saturated zone than the other borings. Visually, the samples were black and degraded. The paper content was around 6% which was the least among all the boreholes. Some of the paper samples collected from the 18.3 m (60 ft.) depth are shown in Figure 4.104 and date back to 2003 (16 years old!). The fine material was higher in the first 3.048 m (10 ft.) of Borehole Z due to the presence of an intermediate cover. The paper fraction was low at all depths. Wood and yard waste were found as 42% in 6.096 m (20 ft.) depth. The main source of wood was from the housing industry and was in good shape. The average composition of Borehole Z was 6% paper, 11% plastic, 3% textile + leather, 16% yard and wood waste, 1% metals, 1% Styrofoam and sponge, and 62% others (mixed other objects and fines).

The average composition of the landfilled samples, including all borings, is presented in Figure 4.105. The major component of the landfilled waste was determined to be degraded fines and soils (60%). Paper and yard/wood waste were both 11%, while plastic was found to be 9% of the total waste mass.

A higher amount of fine content means that most of the materials can be reused as backfill soils, pavement subgrade soils or other construction materials. Other materials (paper/plastic/woods) can be used as fuel for WTE facilities or any other facilities, as they may have a higher calorific value. The effect of degradation on the calorific values of plastics/papers and woods needs to be evaluated.

(a) (b)

Figure 4.104 (a) Waste sample from Borehole Z (b) Paper found at 60ft. depth. *Source*: Courtesy of SWIS.

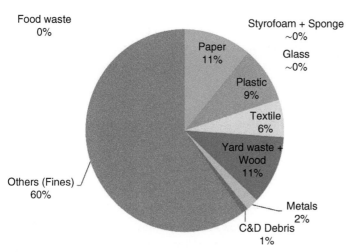

Figure 4.105 Average compositions of waste sample from the City of Irving Landfill.

4.4.3.3.3 Comparison of Fresh and Landfilled Waste Composition

The average composition of the landfilled waste collected from the closed section of the landfill was compared to the annual average of fresh MSW from the working face of the City of Irving's Hunter Ferrell Landfill, as illustrated in Figure 4.106.

As expected, the percentage of paper in the landfilled waste was less than that of the fresh waste, due to degradation of the landfilled waste; the percentage of plastics was also lower. The landfilled waste was approximately 35 years old and plastic packaging was not widely used until the 1990s, which might explain the lower amount of plastics. The amount of soils and fines were higher in the landfilled waste than in the fresh MSW, which might be because of the presence of degraded fines and lot of cover soil mixture, which was absent in the fresh waste samples. As expected, food waste was completely

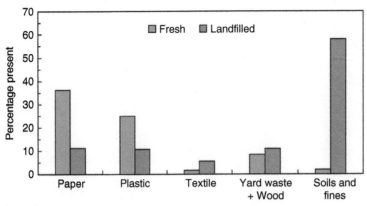

Figure 4.106 Comparison of fresh and landfilled waste.

degraded in the mined/landfilled waste. Very slight differences were exhibited between landfilled and fresh waste for yard/wood waste, metals, glasses, and C&D debris.

4.4.3.3.4 Moisture Content

According to Lu et al. (2002), the moisture content in landfills depends on several interrelated factors, including waste composition, waste type, waste properties, local climatic conditions, landfill operation procedures, gas and leachate collection, and water generation and consumption due to biological processes. The moisture contents of boreholes X, Y, and Z on a wet weight basis are presented in Figure 4.107. The moisture contents (wet weight basis) of the samples from boreholes X, Y, and Z averaged 39.90, 21.63, and 43.86%, respectively. The average moisture content of the Z samples was higher than that of the X and Y samples. The percentage of paper was low in the Z samples as a result of the high moisture content. Waste is anticipated to be more degraded with an increase in depth. According to Landva and Clark (1990), the presence of high organic content in MSW increases the moisture content of the waste. Therefore, with degradation, the moisture content might be reduced; however, no significant trend was found in this landfill that supported the idea.

In 2018, samples of fresh waste were taken from the working faces of the Irving Hunter Ferrell Landfill to study their physical composition and other engineering properties. It was found that the average dry weight-based moisture content was 28.12%, and on a wet weight basis, it was 41.74%. The moisture contents of fresh and landfilled waste were also compared, as shown in Figure 4.108. The plots show a decrease in moisture content in the landfilled waste due to decomposition, supporting the theory that moisture content is reduced with degradation.

4.4.3.3.5 Volatile Solids

Volatile solids (VS) are those that are lost by the ignition of dry solids at 1020°F (550°C) and are believed to be a good indication of the energy left in mined waste. As per Kelly

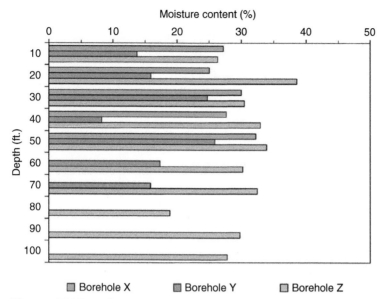

Figure 4.107 Moisture contents of MSW samples from the City of Irving Landfill.

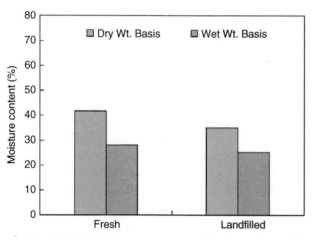

Figure 4.108 Moisture contents of fresh and landfilled waste.

et al. (2006), cellulose could be reasonably predicted from VS, which in turn supports that VS are a good indication of the energy left in the mined waste. In this study, the volatile solids of the mixed waste samples at every depth interval were experimentally found by standard methods APHA Method 2440-E, and the VS test results are illustrated in Figure 4.109 for all of the boreholes. The average volatile solids of mined waste from Irving's landfill was found to be 36.72%; it varied from 15 to 61%, as presented in Figure 4.109.

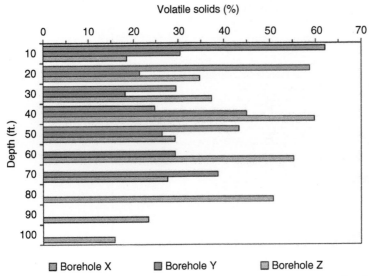

Figure 4.109 Volatile solids at different depths in the City of Irving Landfill.

The average volatile solids of the mixed fresh waste samples from Irving in 2018 was found to be 56.16%. A comparison of the VS with the fresh and landfilled samples is presented in Figure 4.110. Based on the current results, the mined waste experienced 35% more degradation than the fresh waste.

4.4.3.3.6 Calorific Value

Calorific value or the valorization potential is the energy contained in fuel and is determined by measuring the heat produced by the complete combustion of a specified quantity of the fuel. According to Majumder et al. (2008), "calorific value is defined as the amount of heat evolved when a unit weight of the coal/MSW is burnt completely, and the combustion products cooled to a standard temperature of 298 K." It is expressed in British thermal units per pound, i.e. BTU/lb. The calorific value was estimated using a bomb calorimeter. In this study, paper, plastic, wood, and textile were separated from each sample, and representative samples of each were collected and analyzed for their calorific value, using an oxygen bomb calorimeter.

The average calorific values of paper, plastic, yard/wood waste, and textile were found to be 6197, 14 694, 6660, and 7549 BTU/lb, respectively. The calorific values specific to each borehole for the above-mentioned combustible waste types are presented in Figure 4.111. (*1 BTU/lb = 2.33 kJ/kg).

4.4.3.3.7 Comparison of energy potential between fresh and mined solid waste

Another study was conducted at UTA (Ahmed 2019) to evaluate the energy potential of individual combustible fractions (paper, plastic, and yard waste and wood) from fresh

Figure 4.110 Comparison of volatile solids between fresh and landfilled waste.

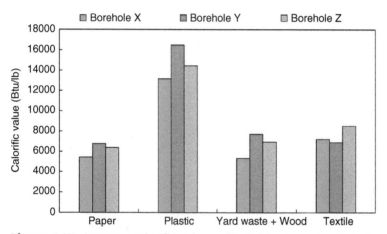

Figure 4.111 Average calorific values of individual combustible fractions. *Note:* 1 BTU/lb = 2.33 kJ/kg.

and mined waste, using a calorimeter, as presented in Figure 4.112. A mined sample was collected from both a conventional landfill and the enhanced leachate recirculation (ELR) landfill, also known as a bioreactor landfill, where generated leachate was recirculated back into the landfill. The mined waste exhibited energy values that are similar to those of fresh waste. Food waste was completely degraded in the mined waste. The overall energy value of the excavated waste was compared with fresh waste from the same landfill and the comparisons are shown in Figure 4.112.

The average calorific values of mixed fresh waste and mixed mined waste are compared in Figure 4.113. It can be seen that the landfilled/mined waste had a lower average than the fresh waste due to degradation. However, if the other waste (mainly fine content) is separated from paper, plastics, wood, and textiles, the calorific value of the mined waste is as high as that of the fresh waste.

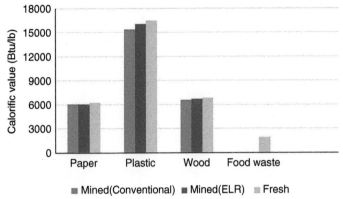

Figure 4.112 Comparison among individual waste fractions. *Note:* 1 BTU/lb = 2.33 kJ/kg.

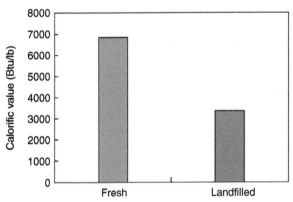

Figure 4.113 Comparison of calorific value of fresh and landfilled waste. *Note:* 1 BTU/lb = 2.33 kJ/kg.

4.4.3.3.8 Factors Affecting Calorific Values

Various parameters, such as depth, landfill operation, moisture content, volatile solids, age of waste, precipitation, and fine fraction were analyzed to understand the behavior of mined solid waste. It was observed that the fine fraction had a decreasing trend with an increase in the calorific value, and the volatile solids had an increasing trend. The moisture content and depth of the landfill did not exhibit any significant correlation with the calorific value. Several factors that impact the calorific value of the mined waste are described below.

- **Effect of Fines**
Fine/degraded material was found to be an excellent indicator of the energy potential of mined waste and was inversely proportional to the energy value.

Fine material increased with a decrease of the energy value, regardless of the depth of the mined sample. Based on the analyses, the excavated waste, which was 0–20% fine, had an average calorific value of 5000–7000 BTU/lb. When the fine material increased up to 40–50%, the energy value decreased to 3000–4000 BTU/lb. Furthermore, the energy potential reduced to 1500–2000 BTU/lb when there was 70–80% fines.

- **Effect of Waste Composition**

Food waste, yard waste, cardboard, and paper fractions are generally considered biodegradable in landfills (Eleazer et al. 1997). The recovery of plastic plays a vital role in assessing the energy potential of excavated waste, and its fraction is usually higher on a weight basis in the higher degraded regions. This is probably because the relative proportions of readily degradable organics, such as paper, cardboard, food, and yard waste decline due to degradation, and the percentage of non-biodegradable fractions corresponding to the overall composition of the waste increase. However, an identifiable trend was not observed in this study, as the waste composition was not uniform in all of the boreholes.

- **Effect of Volatile Solids**

Volatile solids have a direct relation to calorific value, as an increase in VS results in a higher energy yield. Since volatile solids represent the quantity of biodegradable materials present in a waste mass, they can act as an indicator of calorific value. VS can be used as an alternative to the oxygen bomb calorimeter method to determine the energy potential. An approximate energy potential can be estimated based on the relation between VS and calorific values, as shown in Figure 4.114.

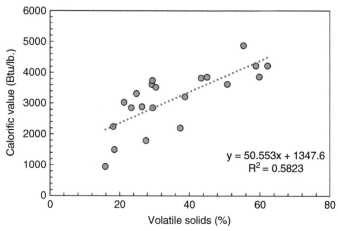

Figure 4.114 Relation between volatile solids and calorific value. *Note:* 1 BTU/lb = 2.33 kJ/kg.

4.4.3.4 Reuse of Mined Biocell Materials

Based on the idea presented in a SMART facility, most of recyclables will be taken out of waste stream before left over waste will be temporarily deposited in biocell. During the biocell operation additives and more moisture will be added to the waste for accelerated decomposition and gas production. Since non-degradable plastics, glass, and metals are removed up front, all of the waste placed in the biocell can be degraded. After the decomposition of the mostly degradable waste component under biocell operation, left over materials are expected to be mostly soil like or compost like materials, most of which can be reused as compost or construction materials based on characteristics study presented in the previous section. The mining of biocells can be completed using a similar process and equipment as described in the case studies.

4.4.4 Waste-to-Energy as a Final Disposal Option

There is no food waste in mined waste; it is comprised of more than 50% fines or other soil-like materials. In developing countries, MSW contains mainly organics and food waste and if the landfill is operated as a biocell, the food wastes degrade very quickly (within 10 years) and the remaining materials are soil like materials. Papers and plastics remain undegraded and can be separated from soil-like materials after digging them out through mining. Once separated, they can either be recycled or used in a WTE facility to generate energy. Based on the research, the calorific value of mined paper/plastics is still very high and can produce energy in a WTE plant. After decomposition of MSW in biocells in developing countries, all of the remaining materials can be reused through recycling and/or energy generation (paper + plastics).

One of the major issues of waste management is the decision-making tool that is needed for help in deciding how to manage the waste that remains after processing. The core of this may depend on the characteristics of the waste, which vary significantly with geographical location; income level; living standard; and socio-economic condition of a country, city, or community. In general, the MSW in developed countries consists of more than 50% paper, plastics, metals and other recyclables, and limited food/organic waste. On the other hand, in developing countries the MSW has more than 50% organic or green waste and less than 20% recyclables. Therefore, the waste management practices are expected to be different in different regions. The appropriate waste management technology for Europe, the U.S., and other developed nations may be completely inappropriate for developing countries and vice versa. One such technology is WTE plants. The main difference between the waste characteristics of developed and developing nations is the moisture content of the solid waste (Figure 4.115). In Europe, the U.S., and other developed nations, the moisture content is usually less than 30%, whereas the moisture content of solid waste in developing countries varies between 60 to 80%.

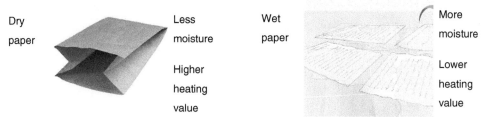

Dry paper	Less moisture	Wet paper	More moisture
	Higher heating value		Lower heating value

Figure 4.115 Dry paper and wet paper.

Therefore, WTE facilities are common in many developed countries, especially in Europe, and they generate heat or electricity for heating the residents' homes. This is a good technology for European countries as the heating value for their waste stream is at the higher end.

Following the success of WTE plants in developed countries and Europe, many developing countries are trying to use them to manage their municipal solid waste. They have had limited success, however, as the moisture content in their waste stream is very high compared to that of developed countries. In some cases, the WTE facilities in developing countries only ran for few days to a few months before they completely shut down the operation. The appropriate choice of technology for managing solid waste is a major issue in developing countries and the wrong one is often selected because of their lack of trained or highly skilled waste professionals. It is, therefore, crucial for us to provide and share guidelines on the various technologies that are available and their selection and use so that developing countries can have successful and sustainable waste management systems.

The calorific or heating values of waste components are the main decision-making tool for managing the waste in a WTE plant. The calorific values of waste components with different moisture levels can be calculated using the following equation:

Lower calorific value of waste,

$$H_{inf} = H_{awf} \times C - 2445 \times W \text{ in kJ/kg} \tag{4.1}$$

where, H_{awf} = Ash and water free calorific value $\approx 20\,000\,\text{kJ/kg}$

C = Combustible matter or volatile solid content (%)

W = Water content (%) (Hoomweg et al. 1999)

Waste for a WTE plant must meet certain basic requirements. The energy content of the waste, the so-called lower calorific value (H_{inf}), must be above a certain minimum level, and the overall composition is also important, as it affects the overall calorific value. Furthermore, in order to operate the plant continuously, the waste generation must be fairly stable throughout the year.

The minimum requirements of waste that is to be incinerated for waste-to-energy generation are shown below (Hoomweg et al. 1999):

- Lower calorific values must be least 7 MJ/kg (7000 kJ/kg) or 6635 BTU/kg
- Moisture content (%W) (< 50%),
- Ash content (%A) (< 60%)
- Combustible matter or volatile solid (%C) (> 25%)

To calculate the overall lower calorific value of a waste mass using Eq. 1, the most accurate way is to characterize the waste by its composition (organic, plastics, cardboard, inerts, and others etc.). The composition influences the overall fuel value significantly. The next step is to determine the water content (%w), and the ash content (%A) in the laboratory. Then the combustible matter (%C) can be measured from the total solid matter and ash content of the individual components of waste. Ash and water-free calorific values (H_{awf}) of the individual components can be found in literature, as these values are mostly constant and do not depend on physical composition, moisture, and ash content. Once all of the parameters are known, the lower calorific value of each component can be calculated using Eq. 1. Finally, from the percentages of the individual components in the waste mass and their corresponding calculated lower calorific values, the overall weighted lower calorific value of the waste mass, can be determined. An example of a sample waste analysis of a developing country, as well as the lower calorific value determined as the weighted average of the heat value of characteristic components of the waste, are presented in Table 4.18.

4.4.4.1 Sample Calculation

Lower calorific value of plastic,

$$H_{inf(plastic)} = H_{awf} \times C - 2445 \times W \ in \ kJ/kg \tag{4.2}$$

where, $H_{awf(plastic)} = 33\,000\,kJ/kg$
 $W = 29\%$
 $A = 7.8\%$

$$C = (100W) - A = 100 - 29 - 7.8 = 63.2\%$$

$$H_{inf(plastic)} = 33000 \times 63.2/100 - 2445 \times 29/100 \ in \ kJ/kg = 20147 kJ/kg$$

4.4.4.2 Lower Calorific Value of Overall Waste Mass

$$= \{70 \times (-817) + 5 \times 20147 + ... + 0.6 \times 2578\} \div 100 = 2120 kJ/kg \ or \ 2.1 MJ/kg$$

The calorific value represents how the food waste component with high moisture content influences the overall calorific value of the waste mass. Therefore, for a country with a high organic waste content in their waste mass, WTE may not be a viable option. Table 4.19 shows the moisture content and lower calorific values of the waste

Table 4.18 Typical lower calorific value of waste mass with high moisture content representing developing country (Calculated based on Hoomweg et al. 1999).

Components	% of waste	Moisture content, W (%)	Total solid, TS (%)	Ash content, A (%)	Combustible Content, C (%)	Calorific value (kJ/kg) H_{awf} (From literature)	H_{inf} (Calculated)
Food	70	80	20	13.3	6.7	17000	−817
Plastic	5	29	71	7.8	63.2	33000	20147
Textile	3.5	33	67	4	63	20000	11793
Paper and cardboard	5	47	53	5.6	47.4	16000	6435
Leather and rubber	1.4	11	89	25.8	63.2	23000	14267
Wood	8	35	65	5.2	59.8	17000	9310
Metals	4.2	6	94	94	0	0	−147
Glass	1.3	3	97	97	0	0	−73
Inerts	1	10	90	90	0	0	−245
Fines	0.6	32	68	45.6	22.4	15000	2578
Weighted average	100	64.5	35.5	8.9	26.6	16822	**2120**

Table 4.19 Moisture content and lower calorific value in different countries.

	Moisture content (%)	Lower calorific value (MJ/kg)	Lower calorific value (BTU/kg)
China (Lu et al. 2017)	52.24	5.16	4893
EU (Manders 2008)	34	9.20	8721
U.S. (Chandrappa and Brown 2012)	19.02	14.70	13 930
Japan (Monbetsu City 2012)	41.88	8.60	8153
South Korea (Seo 2013)	30.29	10.00	9474
Taiwan area (Shu et al. 2006)	49.45	6.02	5703
India (MCD 2004 and TERI 2002)	50	2.93	2777

in different countries, per the literature. Developing countries like India and China have lower heat values due to the presence of large amounts of organic waste and high moisture contents. Among the developed nations, the U.S. has the highest heating value and lowest moisture contents. The heating value is inversely influenced by moisture content, i.e. the higher the moisture content, the lower the heating value.

Due to the variations in waste composition, moisture content, and ash content, heating values vary from country to country, as shown in Table 4.20. Developing countries like Bangladesh and India have more biodegradable components in their waste mass, which leads to a higher moisture content. The ash content is also found to be higher and carbon content is found to be lower in developing countries, which indicates fewer combustible components in the waste mass. Table 4.20 shows that the lower calorific values for Bangladesh and India reported in literature do not meet the minimum criteria for operating a WTE plant. So, the best waste management technique for a developed country may not be the best option for a developing country due to the heterogeneous characteristics of waste.

In addition to the above-mentioned methodology, the calorific value of waste components can also be determined in the laboratory by using a bomb calorimeter.

4.5 SMART FACILITIES CHALLENGES AND OPPORTUNITIES – THE CASE OF ETHIOPIA

SMART facilities as discussed in the previous chapters can address the waste management challenges in developing countries such as Ethiopia by creating business opportunities for the private sector which will be the driving force of the facility.

Table 4.20 Waste-to-energy potential of Bangladesh, India, and USA – Based on average of collected waste.

	Dhaka – Bangladesh	India	USA
Biodegradable	80	60.5	34.3
Recyclable	14	24.1	53.4
Inerts	–	11.0	9
Other	6	4.4	3.3
Ash Content (%)	9.84[a]	15	5.17
Moisture Content (%)	59.42[a]	50	19.02
Lower Calorific Value (BTU/kg)[b]	**673**	**2777**	**13 930**
Source	Islam (2016)	MCD (2004), TERI (2002)	USEPA (2015), Chandrappa and Brown (2012)

[a]Calculated weighted average from composition.
[b]Conversion factor: 1 MJ = 947.817 BTU.

The governance and management of such facilities should be inclusive to accommodate the interest of government, private sector, and civil society organizations (CSO) with environment and social agenda to be successful. The government shoulders the responsibility of managing waste. Consequently, a significant portion of cities budget is allocated to municipal waste management. The challenge is to shift the burden from the city to the private sector by creating an enabling environment for the business to flourish throughout the value chain from collection to processing waste to valuable products which are used by consumers while creating sustainable employment for urban dwellers with a living income. CSOs are concerned of the effect of poorly managed waste on the environment and the adverse effect on human health emanating from air, water, and soil pollution. Hence in the framework of Public Private Partnerships (PPP), the SMART facility should be established to accommodate the interest of the government, the private sector, and the public at large.

4.5.1 Ethiopia Country Profile

i. Ethiopia belongs to the category of low and middle income countries (LMIC).[1] Since 1990, the population doubled to reach nearly 110 million. During the same period, GDP increased sixfold, and GDP per capital tripled. Ethiopia met major targets of the Millennium Development Goals (MDGs): 100% increase in school enrollment; poverty rates reduced by 50%; and life expectancy increased from 47 years to 66 years.

---

1 https://wellcome.ac.uk/grant-funding/guidance/low-and-middle-income-countries.

ii. According to the World Bank,[2] the urban population of Ethiopia reported in the first decade of the century was about 15 million. This is one of the lowest in the world and way below the average for Sub Saharan Africa (37%). It is expected to grow at a rate of 5.4% per year to reach 30% of the population by 2030.

iii. Addis Ababa, the capital city, is home to 25% of the urban population. It is the fastest growing city in Africa in terms of population and economy.[3] The population grew exponentially from 392 000 in 1950 to over three million at a rate of 3.6%. At this rate, the population in 2020 is very likely more than 4.5 million and may surpass six million by 2030.

iv. Addis Ababa hosts the African Union and many international organizations such as Economic Commission for Africa. Its economy is growing at a rate of 14% annually and contributes to nearly 50% the country's GDP.

v. The city is experiencing difficulties resulting from flooding, fire, civil unrest, and displacement. Earthquake, terrorism, and disease outbreak are real threats. Water shortage, power interruption, unemployment, lack of affordable housing, traffic congestion and inadequate waste collection are the major chronic stresses the city is undergoing.[4]

vi. Nearly 75% of the solid waste generated in Addis Ababa is organic. About 17% is recyclable and consist mainly of paper and plastic while the remaining 8% is characterized as hazardous waste. Households generate more than 75% of the waste, followed by public and commercial institutions (18%) and street cleaning which accounts for about 6%.

vii. The City reported about 0.5 kg of waste is generated per person per day and manages to collect 2000 t per day. The estimated population is 4.5 million, thus less than 90% of the waste is collected.

viii. There are nearly 80 associations employing 6000 people to collect household waste house to house. Waste is collected temporarily at designated collection points. Currently there are more than 300 such collecting points across the city. Waste is transferred from the collecting points to the dump site within 24 hours. The city uses 44 compactor trucks for that purpose capable of transferring 55% of the waste to Koshe (also referred as Reppie), the City's dump site, or to the Cambridge waste to energy plant for incineration. Private enterprises transfer the remaining 45% of the waste.

---

2 https://openknowledge.worldbank.org/bitstream/handle/10986/22979/Ethiopia000Urb0ddle0 income0Ethiopia.pdf?sequence = 1&isAllowed = y.
3 https://documents.worldbank.org/curated/en/559781468196153638/pdf/100980-REVISED-WP-PUBLIC-Box394816B-Addis-Ababa-CityStrength-ESpread-S.pdf.
4 http://documents1.worldbank.org/curated/en/559781468196153638/pdf/Addis-Ababa-Enhancing-Urban-Resilience-city-strength-resilient-cities-program.pdf.

ix. The private sector collects and transfers waste from public and private institutions, and international organizations including embassies. There are more than 50 such private enterprises engaged in the business. The City is planning to outsource waste collection and transportation to the private sector fully by 2025.

x. There are nearly 150000 people engaged in street cleaning under the safety net program funded by the World Bank administered by the City's waste management program. Waste is collected by the municipality at curbsides and transported to the dump site.

xi. The City estimates about 65% of the municipality solid waste is organic. About 15% of the waste is recyclable such as paper and plastic and the remaining 20% is composed of hazardous waste from hospitals, industries and other waste including electronics.

xii. The City's municipal solid waste is disposed on open damp site within the city famously known as Koshe or Reppie and has been for more than half a century. There was an effort by the City to close the dump site properly with methane capturing and flaring program which qualified for CDM. At the same time the City built a sanitary landfill outside of Addis Ababa in Oromia region in a place called Sendafa. In addition, the Reppie waste to energy plant was setup at Koshe open dump site to incinerate the waste with the intention to produce 50 MW of electricity.

xiii. The Sendafa landfill was built with financial support by French Development Agency (AfD) and French developers. It met all of the technical requirements for modern sanitary landfill; however, the project failed to undertake a proper social and environmental impact assessment. As the result it became a target during the 2016 civil unrest which forced the City to abandon the project.

4.5.2 SMART Facility in Ethiopia

i. The proposed SMART facility concept can address the City of Addis Ababa's and other cities in Ethiopia solid waste management challenges in conjunction with PPP. The government provides support and incentives for private sector development by establishing industrial zones throughout the country. The SMART facility therefore can be established in the same framework as the industrial zones with similar incentives and privileges to be part of the government program.

ii. The Addis Ababa City has realized waste collection and transportation is more efficient with private sector as compared to the government program. Consequently, there is a plan to outsource waste collection and transportation to the private sector completely by 2025. This undoubtedly will increase waste collection efficiency.

iii. The City's municipal solid waste continues to be disposed of in open dump site which reached its full capacity in 2016. Before the final closure, Addis Ababa city officials were tirelessly looking for a new landfill location and finally found a landfill location just outside the city, Sendafa. There were some major challenges for the Addis Ababa City to dump their waste in an area outside their city and in a different region including:

    a. Placement and finding jobs for the existing waste pickers, who were making a living by picking waste from the Koshe dumpsite. With their operation moved to Sendafa, all of the informal sectors lost their jobs and were therefore under tremendous pressure to find jobs. Therefore, closing the Koshe dumpsite without finding alternative jobs for waste pickers caused a major social unrest among the waste pickers community.

    b. Finding a landfill space in a city like Addis Ababa every 20 years is going to remain a major challenge in foreseeable future. Even after finding a landfill location just outside their city, environmental concern among the people living around the Sendafa landfill created both social and political problems immediately after the start of Sendafa Landfill operation. Finally, the landfill ceased its operation only six months after the starting of their operation. Therefore, waste management SPACE is a major issue for Addis Ababa City.

    c. A SMART facility can help the City of Addis Ababa and other cities in Ethiopia to have a one stop sustainable waste management facility. The waste pickers who were or may be replaced by closure of their open dumps, can be immediately employed in the SMART facility MRF (as they already have experience in waste sorting and collecting recyclables). The major social and economic disaster can be easily diverted through formalizing their jobs at the SMART facility.

iv. In addition, the City is incentivizing recycling with subsidies on organic waste, paper and plastic products. Though this is a good start, to sustain the program it should enact regulations that would encourage waste separation at the source of generation and devise compensation mechanisms by the recycling companies. Sweden's successful recycling program is due to such initiatives.

v. The proposed SMART facility will process the waste collected as described in the previous section. The organic waste will be transformed into compost in the biocells while generating methane. The minimum methane gas generated from a biocell with 3000 t intake capacity per day is more than a million GJ per annum. If managed well that volume can triple.[5] This is the equivalent to displacing

---

5 Based on the calculation made at UTA at different conditions the minimum gas generated for 3000 tons of waste in put per day is slightly more than a million GJ. Under optimum conditions it is close to 3.5 million GJ.

30000–90000 tonnes of charcoal per annum; or 30–90 million liters of kerosene. If methane is sold at the current subsidized price of kerosene ($0.5 per liter or $13 per GJ),[6] the country could save minimum $13 million in foreign currency from oil import. A well-managed biocell could increase the savings to $40 million per year. The biocell therefore will provide sustainable energy for Addis Ababa while reducing the trade balance which is a major concern.

vi. Food waste can be processed with anaerobic digesters to generate additional methane. The potential for biodiesel from waste cooking oil and coffee grounds is high (Haile et al. 2013). This biodiesel can power the waste collection fleet and city buses. The buses in London use B20 biodiesel with 20% mixture from biodiesel derived from coffee ground waste.[7]

vii. Plastic recycling is at early stage in Ethiopia. Plastic pellets are exported to China at the moment. However, the SMART facility can process the pellets to produce building and construction materials as discussed in the previous chapter. Paper recycling is also nonexistent. One of the main reasons is that the supply is not enough for large scale production. The SMART facility can strengthen the supply chain for large scale paper recycling program.

viii. Undoubtedly, the SMART facility will create sustainable employment throughout the value chain from the supply side all the way to the creation of valuable end products. As long as there is waste, employment is guaranteed. It can employ people currently supported by the safety net program at the bottom of the ladder. It can formalize the informal sector such as waste pickers by providing meaningful employment with a living wage. The industry can afford to compensate them handsomely. As the industry expands the demand for waste will increase and recycling will follow the footsteps of Sweden in the framework of circular economy.

ix. Pollution control and mitigation require investment and knowhow. Developing countries like Ethiopia lack both. However, managing waste in the framework of circular economy will generate resources to invest in pollution control and mitigation. The primary objective of the SMART facility is to achieve sustainable waste management. By partnering with center of excellence such as SWIS and networks like C40 and 100 Resilient Cities network of which Addis Ababa is a member, the city can mobilize the resources it needs to achieve and meet the desired objective. Many development partners including the USA, Sweden, UNDP, and the World Bank have a keen interest in supporting such center of excellence.

_____

6 https://www.globalpetrolprices.com/kerosene_prices.
7 https://economictimes.indiatimes.com/news/international/world-news/in-a-first-london-buses-to-be-powered-by-coffee/b20-biofuel/slideshow/61725111.cms.

The SMART facility will address the main challenges of Ethiopia by controlling and mitigating pollution. In doing so it will create sustainable employment with living income. It will create wealth by transforming waste into a valuable resource in the framework of a circular economy. Transforming waste into renewable energy will displace traditional energy use such as charcoal and displace imported oil products, kerosene, and diesel. In doing so, apart from mitigating pollution, it will reduce the trade deficit a major challenge the country is facing.

4.6 TRAINING AND HUMAN CAPACITY BUILDING

According to the International Solid Waste Association (ISWA), one of the major components of successful waste management that is missing in developing countries is the training of personnel employed in their waste management sector. Local authorities and waste management personnel in these countries require essential waste management technology, training, and management assistance. The following section focuses on the need for and importance of training and human capacity building.

During a trip to Accra, Ghana in August 2014 for a training program on landfill and sustainable waste management, Dr. Hossain visited the Tema Landfill in Accra. The site had two very good compactors near the landfill office, but on the working face of the landfill, they were using a backhoe to compact the waste. In conversations with the manager of the landfill, it was revealed that they had the equipment, but no one who could operate it. This is shocking and hard to understand that even though the site had the equipment, they were unable to use it because of they lack trained manpower. This is just one example of the need for trained manpower in developing countries. Others, as well as a few remedial measures, are provided below.

1. During the author's landfill visit at the start of the sanitary landfill in Sendafa, Addis Ababa, he observed that the site had duplicate sets of equipment that had been funded by a donor country for most of their landfill working face operations. However, during my visit in January 2016, most of the equipment was only used for display, as the site did not have any skilled operators, and none of their employees had any formal training on the operation and maintenance of working face equipment and operations, and there were no skilled or trained landfill management staff who could monitor environmental regulations or manage any problems associated with the operations.

2. The landfill operation, especially on the working face, was a complete mess in a few other African, South Asian, Easter European, and Latin American countries visited. They were trying their best, but in most cases, they were doing it wrong.

There was very little or no understanding of leachate management and the implications of leachate mismanagement on the surrounding water bodies and human health. The sites also had very little understanding of health and safety hazards for workers working in waste collection or on the landfill working face. There were absolutely no protocols for managing the waste pickers that were on site, to minimize the chances of their dying. *One landfill manager in Tanzania told the author that "As long as there is not more than one death per week, it's ok. Waste pickers die all the time and its normal."*

3. Most of the city officials in developing countries are ignorant of and/or do not understand the need for a sustainable waste management system in their city. This needs to be changed not only for environment, but also for the health and safety of those living next to the waste management facilities, especially open dumpsites. Their training and knowledge are vital for the overall improvement of their waste management system. They are responsible for allocating the budget and need to understand that training their waste workers will improve their overall health and well-being as well as that of the rest of the community. In some cases, the higher city officials (who are never on the ground, working in waste management) are the only ones who get to attend local and international training sessions. It is good for them to understand what actions need to be taken in their cities for sustainable waste management, but the waste workers, who will be the ones to implement the ideas, also need to be involved in the training program if their waste management program is to be effective and sustainable.

4. The developing world is in desperate need of the technical know-how for establishing and operating a successful waste management system. Therefore, waste management professionals from developing countries need to visit landfills, material recovery facilities, and WTE facilities in developed countries to see first-hand how they operate. Visiting successful waste management facilities in other developing countries with a culture and economic background similar to theirs would also be extremely helpful.

4.6.1 Inception of Solid Waste Institute for Sustainability (SWIS)

The large volumes of waste being generated and the ineffective and unsustainable management approaches that are too often adopted are having an unprecedented impact on the quality of life and the environment in many developing countries. The major cities whose responsibility is to address these issues and create more livable cities are unable to do so in a sustainable way for many different reasons. Two of the main reasons are the lack of understanding of the importance of solid waste management and its impact on health, sanitation, and sustainable cities, and the lack of training for meeting the challenges of an effective and sustainable waste management system.

The existing plethora of national and international institutions and associations are providing useful and effective assistance. However, these countries and their local governments firmly believe that closer collaboration among them on the issue of waste management is needed at this crucial time, to provide more tailor-made, country-driven, and responsive services that can help them address the challenges in a more sustainable way. Some of the middle- and lower-income countries also realize that the dynamic private sector in their countries, while actively participating in the other infrastructure sectors of energy, transport, and water supply, are largely absent in the waste management business. Many developed countries are willing to share their experiences and to more broadly offer their tried and tested solutions but find that an effective regional platform to deliver this exchange is simply absent.

Therefore, it would help tremendously to have a CENTER to address their common problems shared by all cities and provide value-driven solutions. Considering the need for both developed and developing countries, the Organized Research Center of Excellence – Solid Waste Institute for Sustainability (SWIS) was founded in January 2015 at the University of Texas at Arlington (UTA). The local collaborative partners for SWIS are the Cities of Irving, Grand Prairie, Garland, and Lubbock all in Texas, Addis Ababa in Ethiopia, and the international collaborative partner for training and capacity building is the International Solid Waste Association (ISWA).

The mission of the Organized Research Center of Excellence (ORCE) - Solid Waste Institute for Sustainability (SWIS) is to develop clean and healthy urban cities through sustainable waste management.

And the objectives of the ORCE are:

- To research, develop, and implement innovative solutions for sustainable waste management.
- To provide training and support sustainable waste management in both developed and developing countries.
- To help young waste management professionals develop businesses related to waste management and create jobs for the local community.

Some of the specific tasks that SWIS is undertaking are:

- Identifying common operational issues for solid waste management
- Working on the above referenced problems and developing innovative, low-cost, and environmentally sound technological protocols for them and other specific problems.
- Improving the air quality of cities by capturing landfill gases and reducing greenhouse gas emissions.
- Sharing developed protocols with partner cities

- Educating and training landfill owners, operators, and other parties involved in landfill operations
- Bridging the knowledge gap between the requirements of regulatory agencies and solid waste management communities
- Mentoring and assisting young waste management entrepreneurs and creating job opportunities

The SWIS Center combines the expertise of those in science, engineering, and the social sciences to develop new methodologies and solutions for exploring international paths to sustainable waste management for both developed and developing countries.

The training can be in many different formats from workshops and conferences to short courses based on the need of a specific country, region, or locality. The training programs could also be either short-term training (3–5 days), which would target the two groups described below, or a long-term training/diploma program.

a. **Entry level**, for those who are preparing for a career in the solid waste industry – probably recent graduates with little or no background in solid waste management and operation.

b. **Intermediate level**, for persons already employed in the solid waste industry but desire to upgrade their job skills. Their training will focus on introducing new methods, technology, and operation management practices for solid waste. The training for this group may also include information on landfill gas-to-energy programs, an economic/environmental impact assessment of solid waste operations, waste-to-energy, and a cost benefit analyses of different landfill operations, etc.

4.6.2 Training and Educating Solid Waste Professionals – ISWA-SWIS Winter School 2016

The Solid Waste Institute for Sustainability is committed to providing technical support and training for implementing new/advanced technologies for sustainable waste management for the solid waste personnel and young professionals in developed and developing countries. It is working on bridging the gap between developing countries' needs and sustainable waste management solutions. In the long run, SWIS will serve as an international clearinghouse for information on cost-effective and sustainable technologies for sustainable solid waste management systems, including biocell technology (the most advanced form of waste management), landfill gas-to-energy, landfill mining, and a new generation of sustainable/perpetual landfills.

As part of SWIS's mission and long-term goals, the first ISWA-SWIS Winter School was held in January 2016 for an international audience of experienced and emerging solid waste professionals from 27 countries (Figure 4.116). The winter school offered a unique networking opportunity for highly motivated students and waste management

Figure 4.116 ISWA-SWIS Winter School 2016.
Source: Courtesy of SWIS.

professionals; however, the unique aspect of the training was that it was evenly distributed between in-class and hands-on training. It also included laboratory testing and a monitoring program that may be needed during the landfill operation. Since it is affiliated with cities in and around the Dallas Fort Worth (DFW) area in Texas, SWIS has access to the landfills in the area.

As part of the Winter School promotion, SWIS conducted an essay competition for graduate and undergraduate students across the world and received 120 essays from 55 different countries. The essays were extensively reviewed, and 15 students were selected to participate in the Winter School on either a partial or full scholarship.

4.6.2.1 Program Objectives

The objectives set for the ISWA-SWIS Winter School were focused on sustainable waste management. The details of those objectives were the following:

- To involve participants from throughout the world, especially developing countries
- To provide technical expertise in solid waste management
- To provide hands-on experience with landfill operations and equipment

- To promote entrepreneurial opportunities in solid waste management
- To promote knowledge sharing and future collaborations

4.6.2.2 Planned Program Activities to Achieve Goals

The primary goal of the Winter School was to provide a comprehensive overview of sustainable waste management and landfill operations, while maximizing the participants' engagement by exposing them to potential entrepreneurial opportunities in waste management sectors. The following components were designed for and implemented during the Winter School to achieve this goal.

- **Scholarship Opportunities:** SWIS offered scholarships to 15 participants based on an essay competition on sustainable solid waste management. About 120 undergraduate and graduate applicants from more than 50 countries applied for the scholarship. Because of the overwhelming response, the SWIS team hopes to continue providing few scholarships every year. There was also huge response from solid waste personnel and students to attend the Winter School on their own funding, but the SWIS decided to limit the number of participants to 24 since this was the first endeavor. Since the beginning of the 2016 ISWA-SWIS Winter School, setting measurable goals that demonstrate impact has been a high priority.

- **Training Sessions:** The Winter School's curriculum was divided into 27 topics that were covered over a period of two weeks. The order of the sessions was carefully arranged to flow in a manner that made it easy for the participants to follow. The first session introduced the participants to a "Waste Management Overview" and provided them with a definition of waste and its global management perspectives. A session dedicated to entrepreneurial opportunities in waste management was offered, where the participants were introduced to waste economics and business development. On the last day of the Winter School, a panel discussion was held, in which seasoned professionals (Kata Tisza from ISWA, David Biderman from SWANA, Tom Frankiewicz from EPA, Vance Kemler and Brenda Haney from Texas) shared their perspectives of global, national, and local waste management practices (Figure 4.117). The most notable feature of the training sessions was that the 27 sessions were taught by 24 different speakers who were seasoned professionals and academicians and conducted training sessions on their respective expertise.

- **Landfill Site Visit:** Along with the lectures and training sessions, landfill site visits were arranged to provide hands-on experience to the participants. They were trained on specific topics for two days and were taken to the City of Denton landfill every third day to demonstrate the field realities and applications of the material/issues discussed in the training sessions (Figure 4.118). This approach

Figure 4.117 Panel discussions on landfill management on local, national, and global perspectives.
Source: Courtesy of SWIS.

(a)   (b)

Figure 4.118 The City of Denton Landfill visit as part of the training (a) Waste characterization by the participants (b) Working face tour.
Source: Courtesy of SWIS.

was very helpful, since the participants were enthusiastic about the newly learned concepts and engaged in interesting and informative discussions with the landfill personnel. This complemented their classroom learning to a great extent.

- **Hands-on Equipment Operation:** The participants were divided into six groups and were given the opportunity to operate the landfill equipment under the supervision of an experienced equipment operator who ensured their safety (Figure 4.119). This was another unique opportunity for the participants and operating the operating heavy machinery enhanced their interest and engagement in waste management practices.

- **Group Project and Presentation:** The participants were randomly divided into six groups and were assigned a waste management operation (waste collection,

(a)   (b)

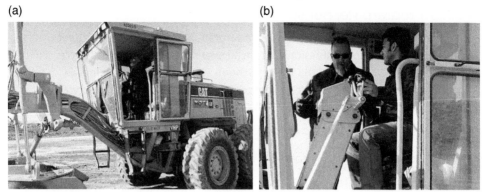

Figure 4.119 Hands-on equipment operation by the participants at the City of Denton Landfill.
Source: Courtesy of SWIS.

Figure 4.120 Group project presentation.
Source: Courtesy of SWIS.

landfill gas-to-energy, etc.) for a group project. Apart from the individual inter-actions and preparations, two sessions of four hours each were arranged to pro-vide them time to share ideas and promote interaction, enjoy group dynamics, and plan future collaborations. On the last day of the Winter School, the groups presented their findings (Figure 4.120). This exercise served a dual purpose: i) it taught the participants to present their ideas convincingly and ii) provided an opportunity for each group to share their acquired knowledge with the others.

4.6.2.3 Program Response

The target audience of the program was undergraduate and graduate students and waste management professionals from all over the world. Since the successful execution of the 2016 ISWA-SWIS Winter School, SWIS has received numerous recognitions from the participants of the school and stakeholders in solid waste management sectors such as ISWA and SWANA in different cities in the world.

"Congratulations on a very successful event. The attendees all learned a lot, were very engaged and motivated, and had fun."
- David Biderman, SWANA CEO

"My experience in ISWA-SWIS winter school is really good. I enjoyed so much. I work for International Solid Waste Association and it is the fourth course of this kind that I managed, not to wreck and it's going really well. I think in comparison with all other we had, it's special. It's more on the operation side and you really get some hands-on training here. So it is really fun that you have day in classroom; afterward you go out into the field and you try to apply some of the knowledge you have gained I love it!!!"
- Kim Winternitz, Event and Project Manager, ISWA

"I think it's awesome. I really like the effort put in there and it's returning. We have a nice mixture of international student and professionals! I am amazed how much people are interested and how much they are enjoying their time!"

- Kata Tisza, Technical Manager, ISWA

"What I like best is the accurate time regulation all through winter school."

- Participant (2016 ISWA-SWIS Winter School

"The winter school was helpful for me. I have learnt a lot technically from the speakers and also different cultures from the participants! The organization was really good, and the organizing team always concerned about helping us!"

- Participant (2016 ISWA-SWIS Winter School)

"I really loved field visits and the laboratory work."
- Participant (2016 ISWA-SWIS Winter School)
"Landfill visit and slope stability is the best."
- Participant (2016 ISWA-SWIS Winter School)

"The winter school was helpful for me, I have learned a lot technically and from other cultures."

- Participant (2016 ISWA-SWIS Winter School)

"Cooperation of the research group with landfill management and networking with specialists all over the world is the key of success!" **- Participant (2016 ISWA- SWIS Winter School)**

"ISWA and SWIS should try to do more to capture the developing countries as the idea of solid waste management we have learnt from winter school will be appreciated more in developing countries!" **- Participant (2016 ISWA-SWIS Winter School)**

"Great fieldtrip." **- Participant (2016 ISWA-SWIS Winter School)**

"Field visits. They were really interesting because we could apply what we learned in class." **- Participant (2016 ISWA-SWIS Winter School)**

"Everything is good better than I expected." **- Participant (2016 ISWA-SWIS Winter School)**

"Excellent." **- Participant (2016 ISWA-SWIS Winter School)**

The ISWA-SWIS Winter School 2016 set a new standard for training programs in the solid waste management sector. The SWIS team successfully completed the 2016 ISWA-SWIS Winter School by educating and training solid waste management personnel involved in landfill operations from Texas, USA, and other developed and developing countries. Due to the overwhelming feedback from the participants, the ISWA Managing Director reached out to SWIS Director, Dr. Sahadat Hossain.

4.6.2.4 Future Continuation of the Program

The achievements of the 2016 ISWA-SWIS Winter School have been inspiring, motivating future possibilities and opportunities for the program. The SWIS team is committed to continuing the program every year, based on the success of the 1st winter school. Every year, participants will be selected from developing countries that did not participate in previous years, as SWIS wants to reach out to more parts of the globe and involve more and more developing countries in sustainable waste management practices.

Figure 4.121 ISWA awards 2019 in Bilbao, Spain.
Source: Courtesy of SWIS.

In January 2020, SWIS completed its fifth successful SWIS training program. It has trained participants from more than 80 countries (including participants from every continent). In 2019, the Winter School program was honored as one of the best global sustainable waste management training programs by ISWA during the annual ISWA Conference Gala Dinner in Bilbao, Spain (Figure 4.121).

The success of the Winter School has been recognized by the participants throughout the five-year run before the pandemic. One of the participants from the Winter School 2020, Paola Nathali Meza, posted a blog in the ISWA website sharing her experience.

Paola Nathali Meza Ramos, originally from Peru, studying at The Autonomous University of Madrid, Spain

ISWA-SWIS offered me the opportunity to participate in the Winter School held at the University of Texas at Arlington. The participants were made up of professionals in waste management from 26 different countries. The diversity of the participants and the shared interest in waste management made us connect immediately. Different realities with a common problem, inadequate waste management.

The program included theoretical classes by leading professionals, experts in the area, as well as group work in specific cases to reinforce the concepts learned throughout the classes.

My group was formed by Abraham from the United Kingdom, Daniana from Colombia, Daniel from the United Kingdom, Eshetu from Ethiopia, Maria from Bangladesh, Mohammed from India, Nasser from Oman, me from Peru and our group leader from Bangladesh. We were in charge of waste management for the city of Bahir Dar, Ethiopia.

The program also included technical visits to various waste management facilities, including City of Irving Landfill, City of Grand Prairie Landfill, City of Garland Landfill and City of Fort Worth-Republic Service.

The program concluded with the presentation of waste management projects for the cities assigned to each group and the awarding of certificates by Dr. Sahadat Hossain and Vance Kemler on behalf of SWIS, and James Law on behalf of ISWA.

Beyond the professional and academic learning, I received, I must highlight the work of the organizers in helping us to form bonds of friendship. I consider it an important factor to stay in touch for a long time and collaborate more easily on future projects that help improve waste management systems around the world.

I finished the two weeks of Winter School with the optimism for the innovative projects that are being carried out. I also ended up with new friends who feel the same passion for waste management and understand the importance it has in an increasingly populated society: my "pack of wolves", as Dr Hossain would say.

I am very grateful to have been chosen to participate in the Winter School, a unique experience that I would definitely recommend to all those interested in waste management.

- *Extracted from www. iswa.org*

REFERENCES

Ahmad, M.S. (2014). Low density polyethylene modified dense graded bituminous macadam. *International Journal of Engineering Trends and Technology 16*: 366–372.

Ahmed, R. (2019). Evaluation and prediction of energy potential of landfill mined solid waste. Dissertation.

Albalak, R. (1997). Cultural practices and exposure to particulate pollution from indoor biomass cooking: effects on respiratory health and nutritional status among the Aymara Indians of the Bolivian highlands. Doctoral dissertation, University of Michigan.

Aurpa, S. S. (2021). Characterization of MSW and plastic waste volume estimation during Covid-19 pandemic. Diss. The University of Texas at Arlington.

Barlaz, M.A., Ham, R.K., and Schaefer, D.M. (1989). Mass-balance analysis of anaerobically decomposed refuse. *Journal of Environmental Engineering 115* (6): 1088–1102.

Barlaz, M.A., Ham, R.K., Schaefer, D.M., and Isaacson, R. (1990). Methane production from municipal refuse: a review of enhancement techniques and microbial dynamics. *Critical Reviews in Environmental Science and Technology 19* (6): 557–584.

Barlaz, M.A., Bareither, C. A., Hossain, A., et al. (2010). Performance of North American bioreactor landfills. II: chemical and biological characteristics. *Journal of Environmental Engineering 136* (8): 839–853.

Behera, D., Dash, S., and Malik, S.K. (1988). Blood carboxyhemoglobin levels following acute exposure to smoke of biomass fuel. *The Indian journal of medical research 88*: 522–524.

Braungart, M. and McDonough, W. (2009). *Cradle to Cradle*. Random House.

Bruce, N., Perez-Padilla, R., and Albalak, R. (2000). Indoor air pollution in developing countries: a major environmental and public health challenge. *Bulletin of the World Health organization 78*: 1078–1092.

Buffière, P., Loisel, D., Bernet, N., and Delgenes, J.P. (2006). Toward new indicators for the prediction of solid waste anaerobic digestion properties. *Water Science and Technology 53* (8): 233–241.

Calma, J. (2020). The COVID-19 pandemic is generating tons of medical waste. *The Verge* (26 March). https://www.theverge.com/2020/3/26/21194647/the-covid-19-pandemic-is-generating-tons-of-medical-waste.

Chandrappa, R. and Brown, J. (2012). *Solid Waste Management: Principles and Practice*. Springer Science & Business Media.

Chen, C.W., Salim, H., Bowders, J.J. et al. (2007). Creep behavior of recycled plastic lumber in slope stabilization applications. *Journal of Materials in Civil Engineering 19* (2): 130–138.

CITYLAB (2019). How American Recycling Is Changing After China's National Sword. *Bloomberg CityLab*. https://www.bloomberg.com/news/articles/2019-04-01/how-china-s-policy-shift-is-changing-u-s-recycling.

Cossu, R., Hogland, W., and Salerni, E. (1996). Landfill mining in Europe and USA. *ISWA Yearbook* 107–114.

Delongui, L., Matuella, M., Núñez, W.P. et al. (2018). Construction and demolition waste parameters for rational pavement design. *Construction and Building Materials 168*: 105–112.

Department of Environment and Natural Resources. 2003. Revised Procedural Manual forDAO 2003–30. Philippines.

Deublein, D. and Steinhauser, A. (2011). *Biogas from Waste and Renewable Resources: An Introduction*. Wiley.

Eder, B. and Schulz, H. (2007). Biogas basis: practice, design. In: *Plant Engineering, Examples and Costs*. Staufen: Ökobuch.

Eleazer, W.E., Odle, W.S., Wang, Y.S., and Barlaz, M.A. (1997). Biodegradability of municipal solid waste components in laboratory-scale landfills. *Environmental Science & Technology 31* (3): 911–917.

Enayetullah, I. and Hashmi, Q.S.I. (2006). Community based solid waste management through public-private-community partnerships: experience of waste concern in Bangladesh. *3R Asia conference*, Tokyo, Japan.

EPA (2010). Landfill Gas Energy Project Development Handbook. Landfill Methane Outreach Program (LMOP), Climate Change Division, U.S. EPA. January 2010.

Faour, A.A., Reinhart, D.R., and You, H. (2007). First-order kinetic gas generation model parameters for wet landfills. *Waste Management 27* (7): 946–953.

Food and Agriculture Organization (FAO) (2011). *Global Food Losses and Food Waste – Extent, Causes, and Prevention*. Rome, Italy: FAO.

Frishberg, H. (2020). Littered masks and gloves filling streets, becoming safety hazard. *NY Post* (21 April). https://nypost.com/2020/04/21/littered-masks-and-gloves-filling-streets-becoming-safety-hazard/.

Gawande, A., Zamare, G., Renge, V.C. et al. (2012). An overview on waste plastic utilization in asphalting of roads. *Journal of Engineering Research and Studies 3* (2): 1–5.

Geyer, R., Jambeck, J.R., and Law, K.L. (2017). Production, use, and fate of all plastics ever made. *Science Advances 3* (7): e1700782.

Global Methane Initiative (2013). *The U.S. Government's Global Methane Initiative Accomplishments Annual Report*. https://www.epa.gov/sites/production/files/2016–01/documents/usg_2013_accomplishments_0.pdf.

Grabar, H. (2013). What If Roads Lasted Twice As Long? An innovation in Texas could extend highway lifespans and decrease repair spending. Bloomberg.com (12 September). https://www.bloomberg.com/news/articles/2013-09-12/what-if-roads-lasted-twice-as-long (accessed 7 February 2022).

Gunaseelan, V.N. (2004). Biochemical methane potential of fruits and vegetable solid waste feedstocks. *Biomass and Bioenergy 26* (4): 389–399.

Gustafsson, J., Cederberg, C., Sonesson, U., & Emanuelsson, A. (2013). The methodology of the FAO study: Global Food Losses and Food Waste-extent, causes and prevention- FAO, 2011.

Haile, M., Asfaw, A., and Asfaw, N. (2013). Investigation of waste coffee ground as a potential raw material for biodiesel production. *International Journal of Renewable Energy Research* 3 (4): 854–860.

Ham, R.K. and Barlaz, M.A. (1987). Measurement and prediction of landfill gas quality and quantity. In: *Sanitary Landfilling: Process, Technology, and Environmental Impact*, 155–166. New York. 1989: Academic Press 4 tab, 6 ref.

Hettiaratchi, J.P.A. (2007). New trends in Waste Management: North American Perspective. *Proceedings of the International Conference on Sustainable Solid Waste Management*, Chennai, India (5–7 September 2007), pp. 9–14.

Hınıslıoğlu, S. and Ağar, E. (2004). Use of waste high density polyethylene as bitumen modifier in asphalt concrete mix. *Materials Letters* 58 (3–4): 267–271.

Hoornweg, D. and Bhada-Tata, P. (2012). *What a Waste: A Global Review of Solid Waste Management*. World Bank.

Hoornweg, D., Thomas, L., and Otten, L. (1999). Composting and its applicability in developing countries. *World Bank Working Paper Series* 8: 1–46.

Hossain, M.S., Gabr, M.A., and Barlaz, M.A. (2003). Relationship of compressibility parameters to municipal solid waste decomposition. *Journal of Geotechnical and Geoenvironmental Engineering 129* (12): 1151–1158.

Hossain, M.S., Sivanesan, Y.S., Samir, S., and Mikolajczyk, L. (2014). Effect of Saline Water on Decomposition and Landfill Gas Generation of Municipal Solid Waste. *Journal of Hazardous, Toxic, and Radioactive Waste 18* (2): 04014002.

Hossain, S., Khan, S., and Kibria, G. (2017). *Sustainable Slope Stabilization using Recycled Plastic Pins*. CRC Press.

Hunte, C. A. (2010). Performance of a full-scale bioreactor landfill (Vol. 72, No. 03). Doctoral dissertation, University of Calgary.

Interstate Technology & Regulatory Council (2006). *Characterization, Design, Construction, and Monitoring Of Bioreactor Landfills. ALT-3*. Washington, DC: Interstate Technology & Regulatory Council, Alternative Landfill Technologies Team. www. itrcweb.org.

Inzerillo, L., Di Mino, G., Bressi, S. et al. (1995). Image Based Modeling Technique for Pavement Distress Surveys: a Specific Application to Rutting. *International Journal of Engineering and Technology* 16 (5): 1–9.

Islam, M.R. (2015). *Thermal Fatigue Damage of Asphalt Pavement*. The University of New Mexico.

Islam, K.M. (2016). Municipal solid waste to energy generation in Bangladesh: possible scenarios to generate renewable electricity in Dhaka and Chittagong city. *Journal of Renewable Energy* 2016.

Karanjekar, R. V. (2013). An improved model for predicting methane emissions from landfills based on rainfall, ambient temperature, and waste composition. Doctoral dissertation, University of Texas at Arlington, Arlington, Texas.

Kelly, R.J., Shearer, B.D., Kim, J. et al. (2006). Relationships between analytical methods utilized as tools in the evaluation of landfill waste stability. *Waste Management 26* (12): 1349–1356.

Khalid, A., Arshad, M., Anjum, M. et al. (2011). The anaerobic digestion of solid organic waste. *Waste Management 31* (8): 1737–1744.

Khan, M. S. (2014). Sustainable slope stabilization using recycled plastic pin in Texas. Doctoral dissertation. The University of Texas at Arlington.

Lampo, R.G. and Nosker, T.J. (1997). *Construction Productivity Advancement Research (CPAR) Program: Development and Testing of Plastic Lumber Materials for Construction Applications.* Champaign IL: Construction Engineering Research Lab (ARMY).

Landva, A.O. and Clark, J.I. (1990). Geotechnics of waste fill. In: *Geotechnics of Waste Fills – Theory and Practice* (ed. G.D. Knowles). ASTM International.

Lardinois, I. and van de Klundert, A. (1993). *Organic Waste: Options for Small-Scale Resource Recovery*, vol. 1. Technology Transfer for Development.

Latif, M. B. (2021). Effect of sludge content on different types of food waste degradation in anaerobic digester. Diss. The University of Texas at Arlington.

Lee, G. F. and Jones-Lee, A. (1999). Unreliability of Predicting Landfill Gas Production Rates and Duration for Closed Subtitle D MSW Landfills. Report of G. Fred Lee & Associates, El Macero, CA.

Lohri, C., Vögeli, Y., Oppliger, A. et al. (2010). Evaluation of biogas sanitation systems in Nepalese prisons. *Water Practice and Technology 5* (4).

Lopez, L., and Kemper, C. 2008. Lessons learned from evaluating source separated materials recovery facilities. *SWANA–Northwest Regional Solid Waste Symposium.* Tampa Bay, Florida (17 April).

Lu, T.H., Qian, X.D., and Guo, Z.P. (2002). An Estimation on Gas Generation Of Municipal Solid Waste Landfill. *Journal of Safety and Environment 6.*

Lu, J.W., Zhang, S., Hai, J., and Lei, M. et al. (2017). Status and perspectives of municipal solid waste incineration in China: a comparison with developed regions. *Waste Management* 69: 170–186.

Majumder, A.K., Jain, R., Banerjee, P., and Barnwal, J.P. (2008). Development of a new proximate analysis-based correlation to predict calorific value of coal. *Fuel 87* (13–14): 3077–3081.

Manders, J., 2008. Life cycle assessment of the treatment of MSW in "the average" European waste-to-energy plant. Confederation of European Waste-to-Energy Plants. http://www.cewep.com/storage/med/media/congress2008/228_LCA_Jan_Manders.pdf.

MCD (2004). Feasibility study and master plan report for optimal solid waste treatment and disposal for the entire state of Delhi based on public and private partnership solution Municipal Corporation of Delhi, Delhi. *India* 2004.

McLaren, M.G. (1995). Recycled plastic lumber and shapes design and specifications. In: *Restructuring: America and Beyond*, 819–833. ASCE.

Modarres, A. and Hamedi, H. (2014). Effect of waste plastic bottles on the stiffness and fatigue properties of modified asphalt mixes. *Materials and Design 61*: 8–15.

Monbetsu City (2012). Basic plan for municipal solid waste treatment and disposal.

Montanelli, E.F. (2013). Fiber/polymeric compound for high modulus polymer modified asphalt (PMA). *Procedia-Social and Behavioral Sciences* 104: 39–48. Chicago.

Picheta, R. (2020). Coronavirus is causing a flurry of plastic waste. Campaigners fear it may be permanent. *Ocean Grants* (4 May). https://www.oceangrants.org/ocean-news/2020/5/4/coronavirus-is-causing-a-flurry-of-plastic-waste-campaigners-fear-it-may-be-permanent.

Plastics Europe (2013). *Plastics – the Facts 2013*. https://www.plasticseurope.org/application/files/7815/1689/9295/2013plastics_the_facts_PubOct2013.pdf.

Pohland, F.G. (1975). Acclerated solid waste stabilization and leachate treatment by leachate recycle through sanitary landfills. *Progress in Water Technology*. 753–765.

Rahman, M.H. (2011). Waste Concern: A Decentralized Community-Based Composting Through Public-Private-Community Partnership.

Rahman, N. (2018). Sustainable waste management through operating landfill as bio-cell. Doctoral dissertation. The University of Texas at Arlington.

Sabina, K.T.A., Sangita, S.D., and Sharma, B.M. (2009). Performance evaluation of waste plastic/polymer modified bituminous concrete mixes. *Journal of Scientific and Industrial Research* 68: 975–979.

Sangal, A. (2020). Discarded masks and gloves are becoming a health hazard as people dump them on street. *CNN* (21 April). https://www.cnn.com/2020/04/21/us/coronavirus-ppe-masks-gloves-environment-hazard-trnd/index.html.

Savage, G.M. (1993). Landfill mining: past and present. *Biocycle* 1993: 58–61.

Seo, Y. (2013). Current MSW management and waste-to-energy status in the Republic of Korea. Master's dissertation, Department of Earth and Environmental. Engineering, Columbia University. www.seas.columbia.edu/earth/wtert/sofos/YS%20Thesis_final_Nov3.pdf.

Shiva, P.K., Manjunath, K.R.K., and Prasad, V.R. (2012). Study on marshal stability properties of BC mix used in road construction by adding waste plastic bottle. *JUSR Journal of Mechanical and Civil Engineering (IOSRJMCE) 2* (2): 12–23.

Hung-Yee Shu, Hsin-Chung Lu, Huan-Jung Fan, Ming-Chin Chang & Jyh-Cherng Chen (2006) Prediction for energy content of Taiwan Municipal Solid Waste using Multilayer Perceptron Neural Networks, Journal of the Air & Waste Management Association, 56:6, 852–858, DOI: https://doi.org/10.1080/10473289.2006.10464497

Smith, K.R., Apte, M.G., Yuqing, M. et al. (1994). Air pollution and the energy ladder in Asian cities. *Energy 19* (5): 587–600.

Taufiq, T. (2010). Characteristics of Fresh Municipal Solid Waste.

Tchobanoglous, G., Theisen, H., and Vigil, S. (1993). *Integrated Solid Waste Management: Engineering Principles and Management Issues*. McGraw-Hill.

Tenenbaum, L. (2020). The amount of plastic waste is surging because of the coronavirus pandemic. *Forbes*. (25 April). https://www.forbes.com/sites/lauratenenbaum/2020/04/25/plastic-waste-during-the-time-of-covid-19/?sh=60ca73457e48.

TERI (2002). *Performance Measurements of Pilot Cities*. New Delhi, India: Tata Energy Research Institute.

Tolaymat, T.M., Green, R.B., Hater, G.R., et al. et al. (2010). Evaluation of landfill gas decay constant for municipal solid waste landfills operated as bioreactors. *Journal of the Air & Waste Management Association* 60 (1): 91–97.

Tsakona, M. and Rucevska, I. (2020). UNEP Baseline report on plastic waste.

U.S. Environmental Protection Agency (1998). National Air Quality and Emissions Trends Report, 1997. https://www.epa.gov/sites/default/files/2017-11/documents/trends_report_1997.pdf.

US EPA (2001). *RCRA Financial Assurance for Closure and Post-Closure*. Audit Report, 2001-P-007, US Environmental Protection Agency, Office of Inspector General, Washington, DC, 30 March 2001.

US EPA (2005). Landfill Gas Emissions Model (LandGEM). Version 3.02 User's Guide, EPA-600/R-05/047.

US EPA (2015). Advancing sustainable materials management: 2013 fact sheet (Assessing trends in material generation, recycling and disposal in the United States).

Vasudevan, R.N.S.K., Velkennedy, R., Sekar, A.R.C., and Sundarakannan, B. (2010). Utilization of waste polymers for flexible pavement and easy disposal of waste polymers. *International Journal of Pavement Research and Technology* 3 (1): 34–42.

Venkat, R.R.V. (2017). Report on the utilization of waste plastic materials in asphalt pavements.

Whitehorse Kids Recycling Book (2020). https://www.whitehorse.vic.gov.au/sites/whitehorse.vic.gov.au/files/assets/documents/Whitehorse%20Kids%20Recycling%20Book-webjpg.pdf (accessed 2 February 2022).

World Health Organization (2000). *Indoor air pollution and household energy*. Health and Environmental Initiative. https://www.who.int/heli/risks/indoorair/indoorair/en.

Xu, C., Shi, W., Hong, J. et al. (2015). Life cycle assessment of food waste-based biogas generation. *Renewable and Sustainable Energy Reviews* 49: 169–177.

Yazdani, R. (2010). *Quantifying Factors Limiting Aerobic Degradation During Aerobic Bioreactor Landfilling and Performance Evaluation of a Landfill-Based Anaerobic Composting Digester for Energy Recovery and Compost Production*. Davis: University of California.

YouTube (2017). Uta researcher uses recycled plastic to strengthen highways. YouTube. https://www.youtube.com/watch?v=c0DResP-F4I&t=6s (accessed 7February 2022).

Zuo, M. (2020). Coronavirus leaves China with mountains of medical waste. *South China Morning Post*. https://sg.news.yahoo.com/coronavirus-leaves-china-mountains-medical-122321916.html.

Chapter 5

Decision Making for Sustainable Waste Management Systems

The Solid Waste Institute for Sustainability (SWIS) is committed to providing technical support and training for implementing new and advanced technologies for sustainable waste management for solid waste personnel and young professionals in developed and developing countries. The idea of creating awareness about "Waste is a Resource and Not a Liability – if waste is properly collected and managed" is one of the major goals of SWIS. The idea of landfilling and putting away

The Waste Crisis: Roadmap for Sustainable Waste Management in Developing Countries,
First Edition. Sahadat Hossain, H. James Law and Araya Asfaw.
© 2022 John Wiley & Sons Ltd. Published 2022 by John Wiley & Sons Ltd.

tremendous amount of resource should be replaced by the idea of Sustainable Resource Management. Therefore, the proposed "**S**ustainable **M**aterials **A**nd **R**esource **T**reatment (**SMART**)" facility addresses the problems of finding space for the construction of new landfills every 20–30 years, spreads knowledge of how collecting generated gas through an anaerobic digester (AD) and biocell and converting it to electricity can provide electricity to people living in remote areas where electric grids are not available, and creates green jobs and provides healthy living conditions for both poor and rich people. This is a perfect example of a circular economy for both developing and developed countries. ***The SMART facility also permanently replaces open dumps and landfills as a sustainable material management facility.***

The sustainable waste management decision flow is categorized as small, medium, and large, based upon the quantity of waste that a specific city generates as presented here:

- ***Small City*** – Waste generation of less than 100 tons/d
- ***Medium City*** – Waste generation between 100 and 500 tons/d
- ***Large/Mega City*** – Waste generation of more than 500 tons/d

Decision making for three different selected cities from three different regions are presented in the next sections. The geographical locations of the selected cities are illustrated in Figure 5.1.

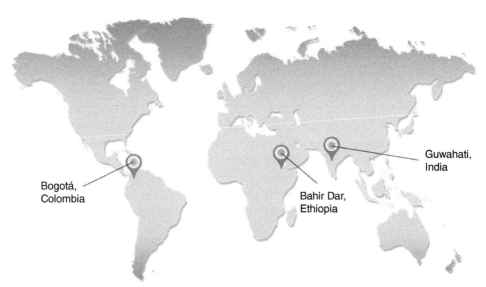

Figure 5.1 Geographical location of the selected cities.

5.1 SMALL CITY – BAHIR DAR, ETHIOPIA

City: Bahir Dar

Country: Ethiopia

Continent: Africa

Population: 168 899 (2016)

Waste Generation: 0.53 kg/capita/day – **90 tons/day** (198 000 lb/day)

Waste Collection Efficiency: 71% (UNEP 2010b)

The city of Bahir Dar is a prominent tourist destination in Ethiopia with a population of 168 899 as per the 2016 census. The total area is 28 km² (11 sq. mi), and the city is located about 578 km (360 mi) from Addis Ababa, which is the capital of Ethiopia. The climate is borderline tropical and the average high and low temperatures are 29 °C (84 °F) and 11 °C (52 °F), respectively. The average rainfall in a year is 1416 mm (55.8 in.).

According to a study conducted by Community Development Research (2011), no formal waste segregation is done at any stage of the waste collection chain in the municipalities of Ethiopia, although some households separate the organic waste for the purpose of producing compost in their garden. The municipality collects solid waste by employing five waste collection companies, known as micro and small enterprises (MSEs), as well as one private company. The five enterprises use hand carts to collect the waste door to door and put it in skip points, where the municipal truck picks it up and takes it to an open dump site located 3 to 4 km (1.86–2.49 miles) from the center of the city. The private company uses hand carts for door-to-door collection and nine trucks for the secondary collection and transportation to the open dump site. The waste collection efficiency of Bahir Dar city was 71% in 2010, according to the UNEP (2010b) study. There is a municipal composting plant in the city with a very small capacity of 0.5 tons/day (1102 lb/d) (Christian 2012). Currently, the city and a private company produce compost from bio-solid waste collected mainly from market centers and hotels and sell it for urban agriculture and to nearby farmers. In 2010, less than 1% of the valuable waste materials were recycled, and that was primarily done at the open dump site by informal recyclers. Plastic bottles are recycled by informal recyclers both before and after collection of the waste. A private company produces high quality charcoal briquettes from the waste for the city. The starting point for any decision making for the sustainable waste management flow or management is determining the waste characteristics.

5.1.1 Waste Characteristics

Based on the report by UNEP (2010a), the waste composition of Bahir Dar city is presented in Table 5.1 and Figure 5.2. The largest proportion of waste is organic (86.6%), and therefore the moisture content of MSW in Bahir Dar city is expected to be very high. The recyclables percentages of MSW is only 7.5%, which is much lower than many other metropolitan cities. Therefore, while decision making for the waste management system, we need to pay special attention to their very high percentages of wet organic waste and absence of source segregation.

5.1.2 Existing Waste Management Practices and Problems for Decision Making

Waste collection is done by the private micro/small enterprises and one private company (this is basically a partnership between public and private organization (PPO). The in-place collection system helps the city with waste collection and reduces

Table 5.1 Waste composition in city of Bahir Dar.

Waste component	Percentage (%)	Amount (tons/d)	Remarks
Organic waste	86.6	78	Biodegradable
Paper	3.3	3	Recyclable
Plastic, Leather	2.2	2	
Textiles	2.2	2	
Metals	0.3	0.5	
Glass	0.6	0.5	Nonrecyclable
Others	4.8	4	
Total	**100**	**90**	

Note: **1 metric ton = 2204 lb.**
Source: Based on UNEP (2010a).

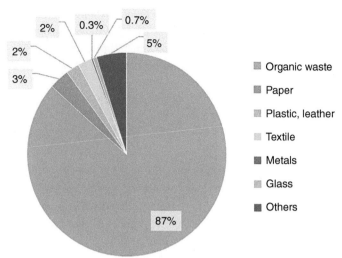

Figure 5.2 Waste composition in city of Bahir Dar.
Source: Based on UNEP (2010a).

the burden on the city for managing and hiring additional staff members for waste collection. A total of 337 waste collectors, 52 hand carts, and 5 trucks are used to collect the waste. However, no waste segregation is practiced and encouraged in the city, which can help the city tremendously to achieve sustainable waste management. Currently, five MSEs and one private company are involved in the waste collection. There are no existing transfer stations for waste sorting and segregation. Major waste management practices now are open dumping, composting, and informal recycling of plastics. There are some major fundamental problems now which include:

1. Scarcity of land for waste management facilities (open dumpsite, sanitary landfills) because of very high cost of available land for sustainable waste management

2. Absence of roadmap to sustainable waste management and the lack of environmental laws or intent to enforce existing minimum environmental laws.

3. Lack of public and policy makers awareness of the need for sustainable waste management and their implications on public health, economy, and environmental safety.

A study by Ejigu (2016) showed that most of the energy consumed by households in Bahir Dar is used for cooking. Firewood and charcoal are both rampantly used as energy sources for cooking, which causes dangerous smoke and serious indoor air pollution. This practice eventually leads to serious health and environmental issues. The need for firewood leads to severe deforestation, and the smoke from burning it produces greenhouse gases. Therefore, having a renewable energy source for Bahir Dar can help the city residents significantly both in terms of their healthy living and better economic condition. Using highly wet organic waste to produce biogas through AD or biocell operation can transform the waste management practices from liabilities to resources for the city and ultimately benefit all stakeholders including city residents, regulators, and the environment.

5.1.3 Proposed Sustainable Waste/Resource Management Approach

The sustainable waste management framework presented in Figure 5.3 shows that creating social awareness is one of the major steps toward sustainable waste management. The city officials can use their authority to use local media (TV/radio/internet) for creating awareness among general public for the need and importance of waste management, collection, source segregation of waste, avoiding littering and throwing their waste into water bodies or around their neighborhood, and ultimately the benefits of proper waste management of their health, well-being and jobs. The regulatory authorities and the general public need to be made aware of the need for proper training for developing, operating, and maintaining the various elements of waste management systems. The training of policy makers is equally if not more important since the implementation of any management approach requires planning and approval from a higher public authority. The integration of all the elements takes time and proper planning but creating awareness among the public and adequate

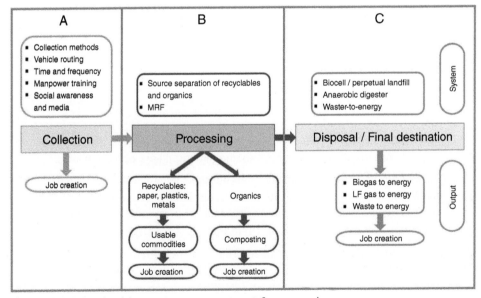

Figure 5.3 Sustainable waste management framework.

training of the workforce need to be the major priorities. Moreover, the government shoulders the responsibility of collecting and managing waste. Consequently, significant portion of cities budget is allocated to municipalities' waste management systems. The challenge is to shift the burden from the city to the private sector by creating an enabling environment for the businesses to flourish throughout the value chain from collection to processing waste to valuable products which are used by consumers while creating sustainable employment for urban dwellers with living income. Hence in the framework of Public Private Partnership (PPP), the SMART facility should be established to accommodate the interest of the government, the private sector, and the public at large.

Based on the waste characteristics of Bahir Dar city, the preliminary sustainable waste management solution for Bahir Dar is outlined below. The system assumes that initially the waste is collected as mixed waste as source separation training and creating awareness can take time.

1. Sort the waste in a material recovery facility (MRF) that is located within a SMART facility.

2. Totally, 86.6% (78 tons [172 000 lb]) of organic waste potentially can be diverted from final disposal

   - Two composting facilities (14 tons [30 864 lb] of capacity each) for handling 28 tons (61 729 lb) of waste

   - Ten small-scale community-level ADs (5 tons [11 023 lb] of capacity each) for handling 50 tons (110 231 lb) of waste (biogas from AD can be used for cooking gas).

3. 8% (7.5 tons [16 534 lb]) of paper, plastic, leather, textiles, and metals can be recycled.

4. The remaining 5.4% (4.5 tons [9920 lb]), along with the remaining reused waste from other methods, can be disposed in a biocell.

5. Possible material diversion = 95%

Since the major waste fraction of city of Bahir Dar is organic waste (86.6%), it will be feasible to operate composting and AD facilities due to the continuous flow of raw materials. The collected mixed waste can be sorted out in a MRF within the SMART facility, where the organic component of the MSW is separated from the recyclables and other components. The following flow diagram illustrates the proposed waste management system for the city (Figure 5.4).

All related activities can take place in one location, a SMART facility. This will make the overall operation manageable and will reduce transportation/hauling times, and costs, eventually reducing the emission of greenhouse gases. Twenty-eight (28) tons (61 729 lb) of the organic MSW will then be used for composting and the remaining organic MSW can be used in small-scale community-level ADs to generate biogas for use in cooking and generating electricity. Ten communities can be equipped with

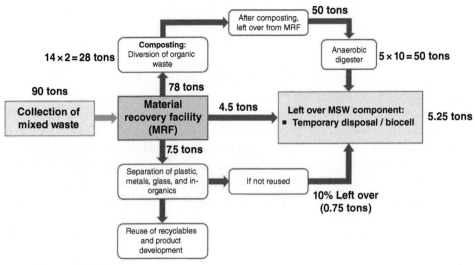

*1 metric ton = 2204 lbs.

Figure 5.4 Proposed waste management system.

underground ADs with a capacity of 5 tons (11 023 lb) each. This will help in alleviating the energy demands, while at the same time promote a sustainable energy alternative. In case of city of Bahir Dar, the AD facilities will be built and operated in different communities, and the community members will be trained and employed there for the operation and management of AD. The remaining waste along with any leftover can be sent to temporary disposal in a biocell. The biocell can be designed to accept any leftover waste from recycling, composting, or AD facilities.

There is potential for job creation in many aspects of waste managements through the SMART facility.

1. The informal sector, which is currently involved in the informal recycling of plastics, can be formalized with permanent jobs due to their experience with recycling

2. More jobs can be created for local residents to work in sorting, composting, and AD facilities within SMART facility.

Jobs in the SMART facility will help local residents making a living out of the waste management facility – the SMART facility will not be viewed as a health hazard or threat for their community (open dumps or landfills are always associated with negative health and environmental concern for the community) rather a source of their income and provider for healthy living conditions. This will ultimately help successful operation of SMART facility in any community in a developing country. The following flow diagram illustrates the proposed SMART facility for city of Bahir Dar (Figure 5.5).

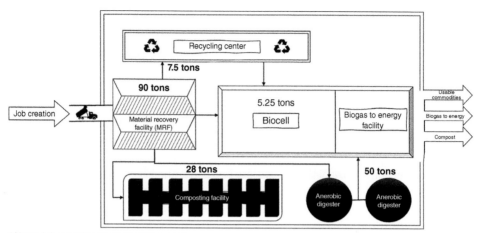

Note: Job creation in each component
Employment of people living around the landfill/open dump to formalize the informal sector
(1 metric ton = 2204 lbs.)

Figure 5.5 SMART facility for city of Bahir Dar.

The MRF-separated recyclables can be reused or recycled to develop new products. There are tremendous opportunities for the young entrepreneurs to get involve in recycling businesses thus creating new sustainable green products, which can meet the local needs and perhaps eventually exported to different countries. The young entrepreneurs can make money for themselves and create more jobs, which will ultimately lead to a clean, healthy city. Eventually with public awareness campaigns and training of city regulators, waste can be segregated at source and sent directly to the community AD facilities for their use, and less organic waste will come to SMART facility. Moreover, the SMART facility can operate in one location for eternity as there is no need for additional space every 20/25 years. This solves a major problem and roadblock for sustainable waste management in developing countries and cities across the world.

5.2 MEDIUM CITY – GUWAHATI, INDIA

City: Guwahati

Country: India

Continent: Asia

Population: 96 2334 (2011)

Waste Generation: 0.461 kg/capita/day – **444 tons/d** (978 852 lb/d)

Waste Collection Efficiency: 90.8% (Singh et al. 2017)

Located between the southern bank of the Brahmaputra River and the foothills of the Shilong plateau, Guwahati is the capital of Assam and the heart of Northeast India. The west and southwest parts of the city are bordered by the Rani rainforest reserve, which gives the area varying high and lowlands. The city had a population of 962 334 in 2011 according to the census, and the total land area of the city is 1598 km² (590 sq. mil). Temperatures vary throughout the year, with an average of 32 °C (89 °F) during the hot season and 12 °C (53 °F) during the cold season. The city experiences about 9.6 months of rainy season, with the month of July receiving the highest rainfall, which averages 330.2 mm (13 in.).

Waste collection in Guwahati is divided into two parts: primary and secondary collection. The primary collection includes house-to-house and commercial establishment waste collection and street sweeping that is done by nongovernment organizations (NGOs) who use tricycles and hydraulic-mounted auto tippers to collect and deposit waste into secondary bins. In the secondary collection, a fleet of modern compactors and tippers operated by the Guwahati Municipal Council (GMC) collects the waste from the secondary collection bins and transports it to two transfer station facilities. The waste is then transferred to the open dumpsite in Boragaon where informal recyclers collect and categorize the recyclable waste. A compost plant with a 50 tons/d capacity that uses wind row composting technology is also located in Boragaon. The starting point for any decision making for the sustainable waste management is determining the waste characteristics.

5.2.1 Waste Characteristics

Based on the report Kalamdhad (2013), the waste composition of Guwahati city is presented in Table 5.2 and Figure 5.6. The largest proportion of the waste is food waste (37.42%). The recyclables percentages is about 42%, which is much higher than many

Table 5.2 Waste composition of Guwahati city (Kalamdhad 2013).

Waste component	Percentage (%)	Amount (tons/d)	Remarks
Food waste	37.42	166	Biodegradable
Yard waste	5.25	23	
Wood scraps	2.45	11	
Paper	16.41	73	Recyclable
Plastic	17.44	77	
Textiles	4.94	22	
Rubber	0.45	2	
Leather	1.97	9	
Metal	0.37	2	
Glass	4.14	18	Nonrecyclable
Others	9.16	41	
Total	**100**	**444**	

Note: 1 metric ton = 2204 lb.

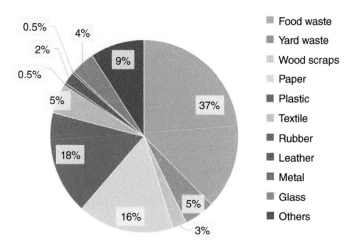

Figure 5.6 Waste composition of Guwahati city (Kalamdhad 2013).

other midsize cities. Therefore, we need to pay special attention to their very high percentages of recyclables and absence of source segregation.

5.2.2 Existing Waste Management Practices and Problems for Decision Making

Waste collection is done by both NGO's and GMC, and their collection efficiency is very high (90%). Waste pickers are the only recyclers for the city who recycle waste from open dumpsite informally. No waste segregation is practiced or encouraged in the city. Ultimately, addressing this can help the city tremendously with better and more sustainable waste management. A compost plant has been installed that has the capacity to produce 50 tons of compost per day in Boragaon, and drum composters have been installed at bulk organic waste generator locations such as vegetable markets, agricultural markets, hotels, and messes. Major waste management practices now are open dumping, composting, and informal recycling of plastics. There are some major fundamental problems now:

- Neighborhood health issues: Health issues abound due to the uncollected waste that is thrown onto the streets, clogging drains, and increasing the number of mosquitoes.

- Infrequent collection: The garbage trucks only operate when dustbins are full (approximately twice a week), and they operate during office hours, which causes a public nuisance.

- Operation of open dumps: Dumping of unprocessed MSW at the open dump-site creates numerous health problems to civilians due to water and air pollution.

- Lack of public and policy makers awareness of the need for sustainable waste management and their implications on public health, economy, and environmental safety.
- Scarcity of land for waste management facilities (open dumpsite, sanitary landfills) because of very high cost of available land for sustainable waste management.
- Absence of roadmap for sustainable waste and lack of environmental laws or intent to enforce existing minimum environmental laws.

A study by Barua (2016) suggests that the community is endangering natural habitats by its open dumpsite operation. The open dumpsite is adjacent to a large natural wetland that is an important destination for migratory birds. The leachate generated in the open dump pollutes the wetland, compromising the lives of innumerable aquatic organisms, migratory birds, and people. Therefore, operating a sustainable waste management system with a biocell and designed bottom liner system mitigate the pollution of the wetland from leachate generated from the open dumpsite.

5.2.3 Proposed Sustainable Waste/Resource Management Approach

The sustainable waste management framework presented in Figure 5.3 shows that creating social awareness is one of the major steps toward sustainable waste management. Clogged drains and mosquitos become major issues for the city residents as the public are throwing trash everywhere around their neighborhoods and littering is widespread in the city. Therefore, the city officials can use their authority to use local media (TV/radio/internet) for creating awareness among the general public for the need and importance of sustainable waste management, collection, source segregation of waste, avoiding littering, and throwing their waste into water bodies or around their neighborhood. All these will ultimately help the public with their health, well-being, and will improve living conditions (as they will be able to get rid of mosquitos with proper waste management and behavior change). The regulatory authorities need to be made aware of more regular and frequent collection of waste from the neighborhood. Otherwise, they become the breeding ground for mosquitos and create a nuisance for the public. Public officials also need to understand that cleaning the city and neighborhood can create a good image for them as a politician or as regulators, which will help them. The need for training waste workers is vital for the success of any waste management program. The training of policy makers is equally if not more important since the implementation of any management approach requires planning and approval from higher public authority. The integration of all the elements takes time and proper planning but creating awareness among the public and adequate training of the workforce need to be the major priorities. Moreover, the government shoulders the responsibility of collecting and managing waste. Consequently, significant portion of cities budget is allocated to municipalities' waste management systems. The challenge is shifting the burden from the city to the private sector by creating an enabling environment for the business to flourish throughout the value chain from collection to processing waste to valuable products which are used by consumers while creating sustainable employment for urban dwellers with living income. Hence, in the framework of PPP, the SMART facility should be established to accommodate the interest of the government, the private sector, and the public at large.

Based on waste characteristics of Guwahati city, the preliminary sustainable waste management solution for Guwahati is outlined below. The system assumes that waste is collected as mixed waste initially as source separation training and creating awareness can take time.

1. Sort the waste in a MRF located within the SMART facility.

    - Divert organic waste from the final disposal: 45% (200 tons) of organic waste. Two composting facilities (50-ton capacity each) for a total of 100 tons of waste

    - Five small-scale AD (ADs) 20-ton capacity each for a total of 100 tons waste (biogas from the AD can be used to generate electricity or cooking gas.)

2. Recycle 42% (185 tons) of paper, plastic, leather, textiles, rubber, and metals.

3. Dispose of the remaining 13% (59 tons), along with the remaining reused waste from other methods to the biocell.

Total waste material diversion potential is: 87% since the major waste fraction of city of Guwahati is organic waste (45%) and recyclables (42%). The collected mixed waste can be sorted out in a MRF within the SMART facility, where the organic component of the MSW is separated from the recyclables and other components. The following flow diagram illustrates the proposed waste management system for the city (Figure 5.7).

The proposed solution is primarily based on the characteristics of the waste and assumes that the collected waste is mixed. Almost half of the MSW waste in Guwahati city is organic (food, wood, and yard waste) that can be processed in composting and AD facilities. The collected mixed waste can be sorted in a MRF where the organic component of the MSW is separated from the recyclables and other components. One hundred (100) tons of the organic waste will be used for composting. Since there is already one composting facility operating with a 50-ton (110231 lb) capacity, the current operation and experience can be beneficial in constructing another composting facility of the same capacity. The compost can be used in the tea gardens of Guwahati city. The remaining organic MSW can be used in small-scale AD to generate biogas that can be used for cooking and/or generating electricity. The 5–20-ton (11023–44092 lb) capacity underground AD can be operated in the SMART facility to generate biogas

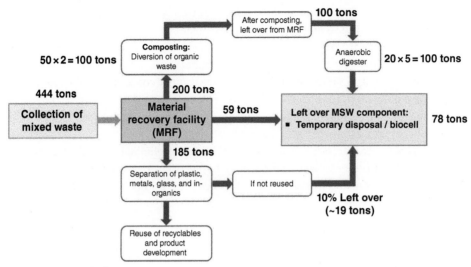

*1 metric ton = 2204 lbs.

Figure 5.7 Proposed waste management system.

for electricity production. The remaining waste can be sent to a biocell for temporary disposal. Any inorganic waste that cannot be reused or recycled can go to a waste-to-energy plant to generate energy by burning if there is sufficient volume and waste with high calorific value to have the needed energy potential for a successful WTE operation. However, this will end the life of the inorganic material and result in an open-ended approach instead of a circular (closed) waste management approach. All of these operations can take place in one location, at the SMART facility as presented here (Figure 5.8). This will make the overall operation manageable, and since all of the processing and disposal will be in the same location, it will reduce transportation/hauling times, manpower/staffing level, and eventually cutting the amount of greenhouse gases.

There is a potential for job creation in many aspects of sections of waste management through the SMART facility.

1. The informal sector, which is currently involved in the informal recycling of plastics, can be formalized with permanent jobs due to their experience with recycling.

2. More jobs can be created for local residents to work in sorting, composting, and AD facilities within SMART facility.

Jobs in the SMART facility will help local residents make a living from the waste management facility – and therefore the SMART facility will not be viewed as a health hazard or threat for their community (open dumps or landfills are always associated with negative health and environmental concern for the community)

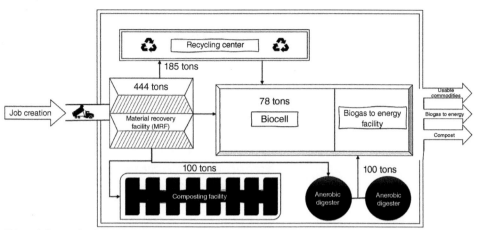

Note: Job creation in each component
Employment of people living around the landfill/open dump to formalize the informal sector
(1 metric ton = 2204 lbs.)

Figure 5.8 SMART facility for Guwahati.

rather a source of their income and provider for healthy living condition. This will ultimately help successful operation of SMART facility in any community in a developing country.

The MRF-separated recyclables can be reused or recycled to develop new products. There are tremendous opportunities for the young entrepreneurs to get involved in a recycling business, creating new sustainable green products, which can meet the local needs and eventually being exported to different countries. The young entrepreneurs can make money for themselves and create more jobs, which will ultimately lead to a clean healthy city. Eventually with an effective public awareness campaign and effective training of city regulators, waste can be segregated at the source of generation and then sent directly to build community AD facilities for their use, and less organic waste will come to the SMART facility. Moreover, SMART facility can operate in one location for a long time as there is no need for additional space every 20/25 years. This solves a major space problem and roadblock for sustainable waste management in developing countries and cities across the world.

5.3 LARGE CITY – BOGOTÁ, COLOMBIA

City: Bogotá
Country: Colombia
Continent: South America

Population: 8 000 000 (2018)

Waste Generation: 0.78 kg/capita/d – **6240 tons/d** (13.8 millions lb)

Waste Collection Efficiency: 95% (UN Environment 2019)

The city of Bogotá is the capital and the largest city in Colombia. It has a land area of about 685 sq. miles and is the most populous city in the country, with an average population growth rate of around 1.4% per year (Vargas Bolivar and Caicedo Moreno 2013). Bogotá city is located in a subtropical highland climate zone where it experiences relatively cool weather. The average temperature ranges from 15 to 20 °C (59–68 °F), and dry and rainy seasons alternate throughout the year, with about 181 days of rain annually. The average rainfall in a year ranges from about 787–1092 mm (31–43 in.).

Bogotá city has an average waste generation rate of 0.78 kg per capita per day, with over a 95% collection rate (UN Environment 2019). Colombia has enforced source segregation of waste, and the categories for separation depend on the waste management plan laid down by the municipality. Bogotá city mandates that recyclables be separated from the rest of the waste. The two general waste management practices are recycling and disposal. The recycling rates are estimated to be over 17% in Colombia (Sarmiento 2017), which can be attributed to the activity of the informal sector and formalized recycling associations. These associations are often small; for example, over 40% of them have fewer than 50 recyclers who collect from various locations such as residences, industries, and commercial centers. Recyclers transport the recyclables to transit points, using hand carts, tricycles, or motorized transports. The most common method of disposal is in sanitary landfills. The utility company is responsible for the collection of nonrecyclables, and they are collected twice weekly and transported in vans with closed doors to avoid problems related to odor. Doña Juana, the landfill in city of Bogotá, has been the principal site for waste disposal for the city since 1989. It implemented a landfill gas capturing project in 2016 to reduce emissions, and approximately 1.7 MW of electricity was generated. The capture and burning of methane have reduced approximately 800 000 tons/year of CO_2 emissions (Biogas Doña Juana 2019). This was an exemplary pilot project for reducing emissions from final disposal sites in Colombia.

5.3.1 Waste Characteristics

Based on the report in Distrital (2018), the waste composition of Bogota city is presented in Table 5.3 and Figure 5.9. The largest proportion of the waste is biodegradable organic waste (66%). The recyclables percentage of MSW is about 30%, which is comparable to other big cities. Therefore, while decision making for the waste management system, we need to pay special attention to their very high percentages of highly organic waste, partial source segregation, and the transportation systems of informal recyclers.

Table 5.3 Waste composition of Bogotá city (Distrital 2018).

Waste component	Percentage (%)	Amount (tons/d)	Remarks
Organic waste	65.48	4086	Biodegradable
Wood	0.71	44	
Plastic	15.55	970	Recyclable
Textiles	5.52	345	
Paper	5.19	324	
Cardboard	2.92	182	
Leather	0.86	54	
Metal	1	62	
Rubber	0.64	40	
Glass	1.17	73	Nonrecyclable
Others	0.96	60	
Total	**100**	**6240**	

Note: 1 metric ton = 2204 lb.

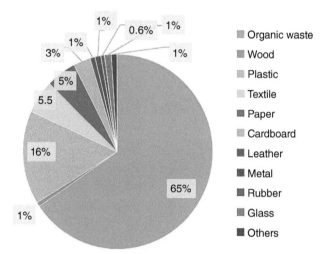

Figure 5.9 Waste composition of Bogotá city (Distrital 2018).

5.3.2 Existing Waste Management Practices and Problems for Decision Making

- Waste collection is done by Bogota city and their collection efficiency is very high (90%). They have both formal and informal recyclers with in-efficient transportation system. Even though, Bogota city mandated source segregation of waste into two categories: recyclables and the others – the source separation is not performed satisfactorily. Therefore, combining the effect of poor transportation systems and poorly implemented source separation, the current recycling rate is about 17%. The most common method of waste management is the Sanitary Landfill, Dona Juana Landfill which has been in operation since 1989 and started capturing methane gas since 2016. However, there are serious operational issues at Dona Juana Landfill, which receives approximately 7000 tons/d (15.4 million lb/day) of solid waste from Bogotá D.C., over the last few years:

- In 2015, the landfill had a major slip failure that was partially due to leachate build-up and improper management, which resulted in over 370 000 tons (0.82 million lb) of waste being spilled from the site (Bnamericas 2015; SSPD 2017). This resulted in an emergency that led severe health impacts in the neighborhood and resulted in severe odors, vectors, and other issues.
- Short remaining life: The currently operating landfill in Bogotá has a short remaining life.
- Waste management tariffs: The waste management tariffs have been increasing significantly and, in some cases, are up to 30% to manage the high costs of collection and disposal (El Tiempo 2018).
- Lack of treatment facilities: There is no treatment facility in Bogotá.
- Financial problems: The entire system of waste management is centralized, which results in high financial stress for a city that has budget constraints.

5.3.3 Proposed Sustainable Waste/Resource Management Approach

Bogotá city is the largest and most populous city in Colombia and has an ever-increasing population growth rate. The city has a congested and busy road network that is focused toward the city center. Furthermore, the sprawling apartments and business centers make the waste collection and transportation process very cumbersome. It is reasonable to have more than one SMART waste management facility in such a large city due to the following benefits:

1. Waste management facilities can be constructed outside the main city at almost equal distances from the city center. This would ensure that the distance that the waste is transported is reasonably short.
2. Less travel time will eventually result in fewer greenhouse gas emissions from the collection trucks.
3. The map of Bogotá City facilitates the division of the city into two sections – North and South.

It is assumed that the north and south sections of the city produce the same amount of waste, which means that each SMART facility will receive the same amount of waste. The city can be divided as shown in Figure 5.10.

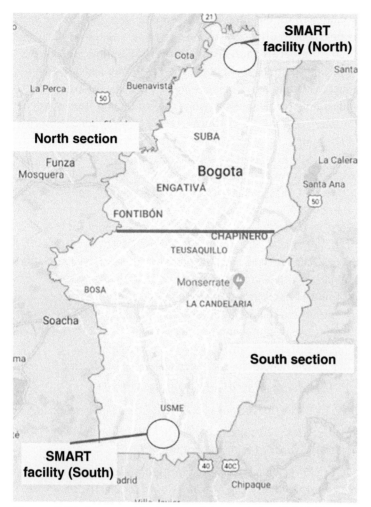

Figure 5.10 Geographical map of Bogotá city divided into north and south sections.

The sustainable waste management framework presented in Figure 5.3 shows that creating social awareness is one of the major steps toward sustainable waste management. The city's percentage recyclables are around 31.5% and has great potential for reuse of recyclables and new products. However, even with city mandated source segregation, the source segregation and recycling rate remained very low. Therefore, the city officials can use their authority to use local media (TV/radio) for creating awareness among general public for their need and importance of waste management, collection, and source segregation of waste. All these will ultimately help public with their health, well-being, and better living. The regulatory authorities need to be made aware of better transportation for recycling and have public private organization (PPO) for encouraging entrepreneurs to participate in waste management business. Public officials also need to understand that cleaning the city and neighborhood can create a good image for them as a politician or regulators, which will help them gain trust of public. The need for training waste workers is vital for the success of any waste management program. The training of policy makers is equally if not more important since the implementation of any management approach requires planning and approval from higher public authority. The integration of all of the elements takes time and proper planning, but creating awareness among the public and adequate training of the workforce need to be the major priorities. Moreover, the government shoulders the responsibility of collecting and managing waste. Consequently, a significant portion of a city's budget is allocated to municipal waste management. The challenge is to shift the burden from the city to the private sector by creating an enabling environment for the businesses to flourish throughout the value chain from collection to processing waste to valuable products which are used by consumers while creating sustainable employment for urban dwellers with living income. Hence, in the framework of PPP the SMART facility should be established to accommodate the interest of government, the private sector, and the public at large.

The preliminary sustainable waste management solution for Bogotá City is outlined below, based on the proposed sustainable waste management for developing countries. Each SMART facility will handle the waste from one half of the city.

1. Sort the waste in an MRF that is located within the SMART facility.

2. Divert organic waste from the final disposal: 66% (2060 tons [4.54 million lb]).
    - Five composting facilities (100-ton capacity each [0.2 million lbs]) for handling a total of 500 tons (1.1 million lb) of waste
    - Fifteen ADs (50-ton [0.1 million lb] capacity each) for handling 750 tons (1.7 million lb) of waste (biogas from AD can be used for electricity generation)

3. Recycle 31.5% (983 tons [2.2 million lb]) of paper, plastic, leather, textiles, rubber, and metals

4. Dispose of the remaining 2.5% (77 tons [0.2 million lb]), along with remaining reused waste from other methods, in the biocell or in a waste-to-energy facility, if feasible.

5. Material diversion is about 68.5%.

Since the major waste fraction of city of Bogota is organic waste (66%) and recyclables (31.5%), the collected mixed waste can be sorted out in a MRF within the SMART facility, where the organic component of the MSW is separated from the recyclables and other components. The following flow diagram illustrates the proposed waste management system for the city (Figure 5.11).

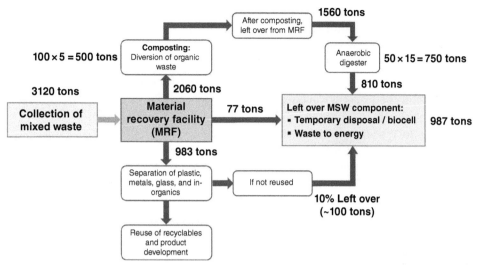

Figure 5.11 Proposed waste management system.
Note: 1 metric ton = 2204 lb.

The proposed solution is primarily based on the characteristics of the waste and assumes that the collected waste is mixed. The mixed waste can be sorted in a MRF within SMART facility, where the organic component will be separated from the recyclables and other components. Five hundred (500) tons of the organic MSW will be used for composting, and 750 tons of organic waste can then be used in the ADs to generate biogas that can be used to generate electricity. The MRF-separated recyclables can be reused or recycled to develop new products. The current rate of recycling (17%) can be considerably increased, and the remaining waste, along with any left over, can be sent for temporary disposal in a biocell. The biocell can be designed to accept any leftover waste from recycling, composting, or the ADs. Since a significant portion of organic waste will be diverted to the biocell, the decomposition rate will be very high, and electricity can be generated from the captured biogas. Landfill mining of the perpetual landfill (biocell) will save a lot of land since the landfill space can be reused many times. The Bogota landfill is almost to its final capacity and may need more space for landfill operation in near future. However, they cannot afford to get new landfill space every 20–25 years, and SMART facility can solve the major space issue for them.

There is the potential for job creation in many aspects or phases of waste management through the SMART facility.

1. The informal sector, which is currently involved in the informal recycling of plastics, can be formalized with permanent jobs using their experience with recycling.

2. More jobs can be created for local residents to work in sorting, composting, AD, and biogas to energy facilities within SMART facility.

Jobs in the SMART facility will help local residents make a living out of the waste management facility and therefore the SMART facility will not be viewed as a health hazard or threat for their community (open dumps or landfills are always associated with negative health and environmental concern for the community) rather a source of their income and provider for healthy living condition. This will ultimately help successful operation of SMART facility in any community in a developing country.

Each SMART facility will follow the layout as shown in Figure 5.12.

The MRF-separated recyclables can be reused or recycled to develop new products. There are tremendous opportunities for the young entrepreneurs to get involved in recycling business, creating new sustainable green products, which can meet the local needs and eventually being exported to different countries. The young entrepreneurs can make money for themselves and create more jobs, which will ultimately lead to a clean, healthy city. Eventually, with an effective public awareness campaign and effective training of city regulators, waste can be segregated at the source of generation and then sent directly to build community AD facilities for their use, and less organic waste will come to SMART facility. Moreover, SMART facility can operate in one location perpetually for a long time as there is no need for additional space every 20/25 years. This solves a major space problem and roadblock for sustainable waste management in developing countries and cities across the world.

Finally, the SMART facility will replace the typical waste management through open dumps or landfills as the sustainable waste management system in every

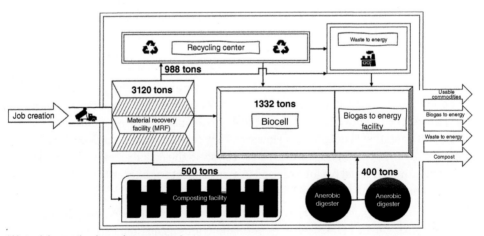

Note: Job creation in each component
    Employment of people living around the landfill/open dump to formalize the informal sector

Figure 5.12 A SMART facility for city of Bogotá.

Note: 1 metric ton = 2204 lb.

community in years to come. A SMART facility's success relies on conducting public outreach and education program, getting alignment with all stakeholders, politicians, and decision-makers, working with waste pickers and creating jobs, and transferring of technologies needed at each stage.

REFERENCES

Barua, V.B. (2016). Open Dumps – Its effect on public life and health. *ISWA-SWIS Proceedings*, pp. 37–44.

Biogas Doña Juana (Mayo de 2019). *Biogas Doña Juana*. Obtenido de http://biogas.com.co/#historia.

Bnamericas (Octubre de 2015). *BNAMERICAS*. Obtenido de https://www.bnamericas.com/en/news/waterandwaste/bogota-gets-fined-for-landfill-mismanagement.

Christian, R.L. (2012). Feasibility Assessment Tool for Urban Anaerobic Digestion in Developing Countries. A participatory multi-criteria assessment from a sustainability perspective applied in Bahir Dar, Ethiopia. MSc thesis in Environmental Sciences (major in Environmental Technology), Wageningen University Netherlands, Final 24 January 2012.

Community Development Research (2011). *Ethiopia Solid Waste & Landfill: Country Profile and Action Plan*. Report produced with funding from the Global Methane Initiative.

Distrital, V. (2018). Veeduria distrital. http://veeduriadistrital.gov.co/sites/default/files/convocatorias/Informe%20de%20Cumplimiento%20Decreto%20371%20de%202010-Vigencias2014-2015_vf.pdf

Ejigu, N. A. (2016). Energy modeling in residential houses: a case study of single-family houses in Bahir Dar City, Ethiopa, Ethiopia. Master thesis, KTH School of Industrial Engineering and Management, Stockholm, Sweden.

El Tiempo (9 de Augusto de 2018). *El Tiempo*. Obtenido de https://www.eltiempo.com/bogota/en-cuanto-quedaran-las-tarifas-de-aseo-en-bogota-para-el-2018-254004.

Kalamdhad, A.S. (2013). *Characteristics of Solid Waste Generated from Guwahati City and Feasibility Solution for Its Management ICRIER's Program on Capacity Building and Knowledge Dissemination on Urbanization in India, Assam Administrative Staff College Guwahati*. Assam: Indian Council for Research on International Economic Relations (ICRIER) 2013.

Sarmiento, M.C. (2017). *Solid Waste Management in Colombia*. Dwaste.

Singh, K.R., Jain, M.S., Goel, G., and Kalamdhad, A. (2017). *Municipal Solid Waste Management in Guwahati City, India*. Guwahati, Assam, India: Indian Institute of Technology.

SSPD (2017). *Informe de Disposición Final de Residuos Sólidos – 2017*. Colombia: Superintendencia de Servivios Públicos Domiciliarios.

UN Environment (2019). *Waste Management Outlook for Latin America and The Caribbean*. UN Environment.

UNEP (2010a). *Assessment of the Solid Waste Management System in Bahir Dar Town and the Gaps identified for the Development of an ISWM Plan*. Forum for Environment, June 2010.

UNEP (2010b). *Solid Waste Characterization and Quantification of Bahir Dar City for the Development of an ISWM Plan*. Forum for Environment, June 2010.

Vargas Bolívar, F. E., & Caicedo Moreno, M. X. (2013). Efectos sobre los precios de la vivienda por su proximidad a un centro comercial en Bogotá.

Chapter 6
Summary

The questions raised in presenting various snapshots in Chapter 1 need appropriate responses and answers, in conjunction with the information embedded in various chapters that urgently needs immediate action plan any given situations and challenges. The answers are presented as follows:

CRISIS 1 AND CRISIS 3
Illegal and Inappropriate Dumping of Waste

<u>Questions</u>: But how can we stop this unhealthy and unsustainable social practice?

Creating social awareness among both private citizens and public authorities are vital and important part of sustainable waste management and creating healthy urban cities. The

The Waste Crisis: Roadmap for Sustainable Waste Management in Developing Countries,
First Edition. Sahadat Hossain, H. James Law and Araya Asfaw.
© 2022 John Wiley & Sons Ltd. Published 2022 by John Wiley & Sons Ltd.

mindset of "out of sight out of mind" or "throw it everywhere and it will go somewhere" or "throw it away and the local government will take care of it" need to change through massive media campaign as described in sustainable waste management framework – Part A. We need to utilize every possible media to create social awareness among mass population on littering and illegal dumping of waste, and their negative consequence on their health, safety, and well-being. Also, we need to inform everyone about potential green jobs associated with increased collection, sorting and separation, reuse of recycling waste components and access to energy for rural setting and associated economic bounty on their life.

Questions: Why are the authorities not taking actions?

In many cases, public authorities are not aware of the negative consequence of littering, illegal dumping on public health and the environment. With proper training and awareness campaign, this can be changed. The benefits of sustainable waste management systems, changing the responsibility and burden from their shoulder to private organizations through PPP need to be explained to them clearly. Once the public officials understand that waste is not a liability but a resource, their mindset about waste management will change and they may allocate more budget and trained manpower for creating healthy urban cities. Because by increasing waste collection, reducing littering and cleaning urban cities, they are not only helping the environment and creating healthy urban cities but also helping themselves. By improving public health conditions through cleaner cities, their image will get a major boast, which can help them personally in the long run.

CRISIS 2

Flash Flooding in Urban Cities

Questions: Why is this unusual flooding happening in the city?

Urbanization, an increase in GDP, and people's purchasing power have caused an exponential increase in consumption and waste generation, in particular, the plastic waste. However, in developing countries, rate of waste collection is between 20 and 60%. Uncollected waste dumped in open places, on roads and streets, in water bodies, and in most public areas is getting into city's drainage system. Moreover, the presence of highly nondegradable plastic bottles and plastic bags is clogging the city's drainage system and causing flash flooding. This is hindering the city's objective to become a world class urban city and can also deter the interest of foreign investors and create a bad image for the city and country as a whole. Sustainable waste management (including increase in waste collection) is therefore vital for city's future development.

- Creating social awareness among both the general public and public authorities on the importance of increasing waste collection.
- Reducing/banning one-time use plastic bags and providing alternative solutions for plastics or plastic bags. Encouraging the use of reusable bags for grocery and other applications.
- Increasing waste (including plastic) collection and proper management.
- Discouraging littering through strict regulations and enforcement of the regulations.

CRISIS 4
Safety for Waste Pickers at Working Face of a Dumpsite

Questions: How can we improve this deadly/unsustainable working conditions for the waste pickers?

- The chaotic scene at the working face can be completely avoided through the implementation of a SMART facility.
- Training and formalizing waste pickers job by employing them in the SMART facility. Health and safety training should also be provided routinely to these formalized waste picker employees.
- Creating social awareness among both the general public and public authorities is vital and an important part of sustainable waste management and creating healthy urban cities.

CRISIS 5
Lebanon Crisis and Issue of SPACE

Questions: How can we solve the major problem with SPACE?

- Implementing a SMART facility and reusing biocell space repeatedly can solve the major roadblock of SPACE for sustainable waste management in many cities around the world. Most cities will not need to look for a new landfill space every 15/20 years if they implemented the SMART facility operation.
- Implementing circular economy through the proposed resource management system and creating jobs for local residents living around the SMART facility.

- Developing human resource capacity by training SMART facility operators and workers.
- Training waste management regulatory agencies about the need for a SMART facility and their implications on the environment, public health, and economic development will be vital for the successful implementation of the SMART facility.

CRISIS 6
Failure of Open Dumpsite in Koshe, Ethiopia

Questions: How to avoid catastrophic failure like Koshe dumpsite? What is the best way to close the existing dumpsite without causing the sociopolitical problems or creating a situation like Lebanon in 2015? Can the waste management facility operate in one place perpetually for a long time?

- Training waste management professional on how to operate existing open dumps and creating sustainable/safe working face operation.
- Training waste management regulatory agencies/authorities on need for a buffer zone between the designed toe of the open dump and people living next to the open dumps. Also, making sure the buffer zone is both created and strictly enforced during the operational period.
- Training operators on safe and efficient operation on compaction of waste at the working face and using stable slope angles during working face operation.
- Developing human resource capacity by training SMART facility operators and workers.
- Training waste management regulatory agencies/authorities about the need for a SMART facility and their implications on the environment, public health, and economic development will be vital for the successful implementation of the SMART facility.

CRISIS 7 AND CRISIS 8
Waste Workers Living Near Dumpsite and Their Health

Questions: How can we improve their lives or livelihoods?

- Waste workers' lives, living on dumpsite or near dumpsite, can be completely changed by implementing a SMART facility and creating jobs for them.

- Training and formalizing waste pickers' job by employing them in a SMART facility. Health and safety training should also be provided to waste pickers for them to understand the importance of raising their children in safe and healthy condition.

- Waste workers are doing a great job for the society and creating healthy urban cities around the globe. They deserve to have healthy/sanitary living conditions for them and their children. Developing cost-effective housing for waste workers around a SMART facility and creating parks or recreational open spaces for their children will make a significant difference for their health and economic condition.

CRISIS 9
Implications of China Ban

<u>Questions</u>: How can we manage plastic waste sustainably? How can we create local market for recyclables and avoid sending plastic waste to landfills to continue in circular economy path?

- The government shoulders the responsibility of collecting and managing waste. Consequently, a significant portion of cities budget is allocated to municipalities' waste management program. The challenge is to shift the burden from the city to the private sector by creating an enabling environment for the business to flourish throughout the value chain from collection to processing waste to valuable products, which are used by consumers while creating sustainable employment for urban dwellers with living income. Therefore, in the framework of Public Private Partnership (PPP), the SMART facility should be established to accommodate the interest of the government, the private sector, and the public at large. Implementing a SMART facility and sorting the mixed waste into different waste components through MRF should be a priority.

- The MRF-separated recyclables can be reused or recycled to develop new products. There are tremendous opportunities for the young entrepreneurs to get involve in recycling business, creating new sustainable green products, which can meet the local need and eventually exported them to different countries. Reusing the recyclables plastics and other waste components will keep them out of the landfill and increase the capacity of the landfill space, and most importantly it will create circular economy.

- The young entrepreneurs can make money for themselves and create jobs for others, which will ultimately lead to a clean healthy city. Moreover, a SMART facility can operate in one location perpetually for a long time as there is no need for additional space every 20/25 years. This solves a major space problem and a roadblock for sustainable waste management in developing countries and cities across the world.

- Therefore, a SMART facility has the potential of creating of jobs for people living around the waste management facilities, making it a source of income rather than a source of distress, sickness, and economic hardship.

Index

A

Addis Ababa, Ethiopia
   current waste management
      situation, 290–291
   general information, 290
   Reppie Waste-to-Energy Plant,
      62–65
   SMART facility proposal, 291–293
   waste situation, 64–65
   see also Koshe (Reppie) open dumpsite;
      Sendafa landfill
ADs *see* anaerobic digesters
air pollution
   dioxins, 1, 35–36, 47, 192
   household use of biomass fuels,
      223–224, 317
   open dumpsites/landfills, 44–47
   open waste burning, 3
   uncontrolled fires at dumpsites, 101
   waste-to-energy plants, 61
anaerobic digesters (ADs)
   biogas generation, 225, 226–227
   optimal waste combination, 228–232
   process, 225–226
   proposals for Bahir Dar, Ethiopia,
      318, 319–320
   proposals for Bogotá, Colombia,
      332, 333, 334
   proposals for Guwahati, India, 325–327
   SMART facility proposal in Ethiopia, 293

   sustainable waste management system,
      176, 222–232
authorities *see* public authorities
automated/advanced waste
      collection vehicles, 24, 25,
      27, 28, 29, 185
automated sorting, material recovery
      facilities, 189

B

Bahir Dar, Ethiopia
   decision making for sustainable waste
      management, 315–321
   existing waste management practices
      and problems, 316–318
   household cooking fuels, 317–318, 319
   waste characteristics, 316, 317
Bangladesh, 71–77
   composting, 72, 220–222
   final disposal, 74
   general information, 71
   major waste problems and solutions,
      34, 36, 60, 74–76
   recycling, 71–73
   waste collection, 71–72, 73
   waste management summary,
      76–77
   waste-to-energy projects, 60–61
Beirut, Lebanon, waste crisis, 5–6,
      41, 43, 112

building material recovery (BMR)
  facilities, 247, 248
burning waste
  illegal burning of plastics at recycling
      factories, 192
  uncontrolled fires at dumpsites,
      46, 48, 101, 161
  see also open waste burning;
      waste-to-energy

C

calorific values
  definition, 280
  fresh versus landfill mined
      waste, 280–283
  fuel sources, 226
  suitability of waste forwaste-to-energy,
      284, 285–288
  waste components, 280
carbon dioxide
  consequences of open waste
      burning, 36
  deforestation and charcoal burning in
      Ethiopia, 62–63
  landfill gas, 40, 44, 240, 241, 269
  see also greenhouse gases
charcoal
  methane fuel comparison, 293
  problems as household
      cooking fuel, 317
  production from waste, 94, 316
  production from wood, 62–63
children
  health impacts from dumpsites,
      9, 54, 135
  indoor air pollution from biomass
      fuels, 223–224
  social awareness of health hazards,
      9, 183

waste pickers' living conditions, 8, 341
working and playing on or near
    dumpsites, 8–9
China
  waste import restrictions and ban, 58,
      190–192, 341
  Wuhan epidemic, 193–194
circular economies
  Colombia, 88
  Ethiopia, 293–294
  material recovery and reuse, 40,
      189–190, 204
  SMART facility, 178, 293, 314, 339, 341
  USA, 204–205, 211
citizen participation see public awareness
    and participation
city officials see public authorities
civil engineering, recycled plastic
    bottles, 204–205
civil unrest see protests
clean material recovery facilities,
    181, 182, 189
closed landfills and dumpsites
  composition and state of
      decomposition, 259–271
  effects of closure on cities, 5–6, 7,
      49, 56, 292
  possibility of biogas production,
      258–259
  post-closure activities, 235
closed loop systems
  sustainable waste management
      systems, 178, 179, 256–258
  see also circular economies
Colombia, 84–89
  Bogotá, 84–85, 86–87, 88–89, 327–335
  final disposal, 86
  general information, 84
  major waste problems and solutions,
      86–88

impact of COVID-19 related plastic
waste, 202
increasing moisture content enhances
decomposition and biogas
production, 234–235
lack of trained operators for
equipment, 294
loss of materials and renewable energy
potential, 40
low moisture content slows
decomposition, 234
management failings, 48, 295
need to minimize waste and increase
capacity of existing sites, 41
post-closure activities, 235
problems for all landfill methods,
39–44
social acceptability, 11, 43, 174–175,
191, 192
traditional landfills compared to
bioreactor landfills and biocells, 236
US regulations, 234
see also open dumpsites/landfills;
sanitary landfills
landslides *see* slope and slip failures
leachates
causing slip failure in landfills, 87, 330
drained in FUKUOKA landfill method, 95
enhanced leachate recirculation
landfills, 234–235, 259, 281, 282
lack of understanding of leachate
management in developing
countries, 295
minimized and contained in sanitary/
engineered landfills, 155, 234
recirculated in bioreactor/ELR
landfills, 234–235, 239, 246–247,
259, 281, 282
unlined landfills in USA, 258
water pollution from open dumpsites, 9,
44, 45, 109, 110, 135–136, 161, 324

Lebanon, 112–118
Beirut waste crisis, 5–6, 41, 43,
112, 115–116
composting, 114–115
final disposal, 115–116
general information, 112–113
major waste problems and solutions,
41, 43, 116–117
open dumpsites/landfills, 41, 115, 116
recycling, 113–114
sanitary landfill at Naameh, 5, 41,
112, 115–116
waste collection, 113
waste management summary, 117–118
LFG *see* landfill gas
littering
COVID-19 effects, 198–200
see also open waste dumping

M

Malaysia, 191, 192
manual sorting, material recovery
facilities, 80, 188, 189, 248
manual waste collection and compaction,
24–26, 27, 28
mass media (TV/newspapers), social
awareness campaigns, 76, 183
material consumption, resources lost in
landfill puts pressure on virgin
materials, 40–41
material flow, sustainable waste
management systems,
180, 181–182
material recovery, *see also* landfill
mining; recycling
material recovery facilities (MRFs)
absent in Ethiopia, 94
automated sorting, 189
Brazil, 80
City of Denton Landfill, 246, 247, 248

waste workers
  improving lives and livelihoods,
    340–341
  need for training in sustainable waste
    management, 295
  trained collectors in developed
    countries, 24
  untrained collectors in developing
    countries, 30–31
water pollution
  bottom ash from waste-to-energy
    plants, 60–61

consequences of open dumping, 33
leachate from open dumpsites/
  landfills, 9, 44, 45, 109, 110,
  135–136, 161, 324
Winter School, ISWA-SWIS Winter School
  XV-XVI, 69, 297–305
WTE *see* waste-to-energy

Z

zoning, subdividing cities for waste
  collection, 186–187